AF411319

Cocrystal Applications in Drug Delivery

Cocrystal Applications in Drug Delivery

Editor

Andrea Erxleben

MDPI • Basel • Beijing • Wuhan • Barcelona • Belgrade • Manchester • Tokyo • Cluj • Tianjin

Editor
Andrea Erxleben
National University of Ireland Galway
Ireland

Editorial Office
MDPI
St. Alban-Anlage 66
4052 Basel, Switzerland

This is a reprint of articles from the Special Issue published online in the open access journal *Pharmaceutics* (ISSN 1999-4923) (available at: https://www.mdpi.com/journal/pharmaceutics/special_issues/cocrystal_applications).

For citation purposes, cite each article independently as indicated on the article page online and as indicated below:

LastName, A.A.; LastName, B.B.; LastName, C.C. Article Title. *Journal Name* **Year**, *Volume Number*, Page Range.

ISBN 978-3-03943-983-6 (Hbk)
ISBN 978-3-03943-984-3 (PDF)

Contents

About the Editor . . . . **vii**

Andrea Erxleben
Cocrystal Applications in Drug Delivery
Reprinted from: *Pharmaceutics* **2020**, *12*, 834, doi:10.3390/pharmaceutics12090834 . . . . **1**

Ricardo Machado Cruz, Tereza Boleslavská, Josef Beránek, Eszter Tieger, Brendan Twamley, Maria Jose Santos-Martinez, Ondřej Dammer and Lidia Tajber
Identification and Pharmaceutical Characterization of a New Itraconazole Terephthalic
Acid Cocrystal
Reprinted from: *Pharmaceutics* **2020**, *12*, 741, doi:10.3390/pharmaceutics12080741 . . . . **5**

Ilma Nugrahani, Rizka A. Kumalasari, Winni N. Auli, Ayano Horikawa and Hidehiro Uekusa
Salt Cocrystal of Diclofenac Sodium-L-Proline: Structural, Pseudopolymorphism, and
Pharmaceutics Performance Study
Reprinted from: *Pharmaceutics* **2020**, *12*, 690, doi:10.3390/pharmaceutics12070690 . . . . **23**

Aneta Wróblewska, JustynaŚniechowska, Sławomir Kaźmierski, Ewelina Wielgus, Grzegorz D. Bujacz, Grzegorz Mlostoń, Arkadiusz Chworos, Justyna Suwara and Marek J. Potrzebowski
Application of 1-Hydroxy-4,5-Dimethyl-Imidazole 3-Oxide as Coformer in Formation of
Pharmaceutical Cocrystals
Reprinted from: *Pharmaceutics* **2020**, *12*, 359, doi:10.3390/pharmaceutics12040359 . . . . **51**

Xavier Buol, Koen Robeyns, Camila Caro Garrido, Nikolay Tumanov, Laurent Collard, Johan Wouters and Tom Leyssens
Improving Nefiracetam Dissolution and Solubility Behavior Using a
Cocrystallization Approach
Reprinted from: *Pharmaceutics* **2020**, *12*, 653, doi:10.3390/pharmaceutics12070653 . . . . **73**

Reynaldo Salas-Zúñiga, Christian Rodríguez-Ruiz, Herbert Höpfl, Hugo Morales-Rojas, Obdulia Sánchez-Guadarrama, Patricia Rodríguez-Cuamatzi and Dea Herrera-Ruiz
Dissolution Advantage of Nitazoxanide Cocrystals in the Presence of Cellulosic Polymers
Reprinted from: *Pharmaceutics* **2020**, *12*, 23, doi:10.3390/pharmaceutics12010023 . . . . **89**

Dnyaneshwar P. Kale, Vibha Puri, Amit Kumar, Navin Kumar and Arvind K. Bansal
The Role of Cocrystallization-Mediated Altered Crystallographic Properties on the Tabletability
of Rivaroxaban and Malonic Acid
Reprinted from: *Pharmaceutics* **2020**, *12*, 546, doi:10.3390/pharmaceutics12060546 . . . . **107**

Bwalya A. Witika, Vincent J. Smith and Roderick B. Walker
A Comparative Study of the Effect of Different Stabilizers on the Critical Quality Attributes of
Self-Assembling Nano Co-Crystals
Reprinted from: *Pharmaceutics* **2020**, *12*, 182, doi:10.3390/pharmaceutics12020182 . . . . **129**

Bwalya A. Witika, Vincent J. Smith and Roderick B. Walker
Quality by Design Optimization of Cold Sonochemical Synthesis of Zidovudine-Lamivudine
Nanosuspensions
Reprinted from: *Pharmaceutics* **2020**, *12*, 367, doi:10.3390/pharmaceutics12090834 . . . . **145**

About the Editor

Andrea Erxleben obtained her PhD in Chemistry from the University of Dortmund (now TU Dortmund) in 1995. Following postdoctoral work with Professor Jik Chin at McGill University, Montreal, Canada, she returned to Dortmund where she completed her "habilitation" and obtained the "venia legendi" for Inorganic Chemistry in 2002. She moved to the National University of Ireland Galway where she is now a Senior Lecturer. She held guest professorships at the University of Vienna and is a member of the Editorial Board of *Pharmaceutics* and the Editorial Advisory Board of *Crystal Growth and Design*. Her research interests are in the fields of Medicinal Inorganic Chemistry and Pharmaceutical Solid-State Chemistry.

Editorial

Cocrystal Applications in Drug Delivery

Andrea Erxleben [1,2]

[1] School of Chemistry, National University of Ireland, H91TK33 Galway, Ireland;
andrea.erxleben@nuigalway.ie

[2] Synthesis and Solid State Pharmaceutical Centre (SSPC), V94T9PX Limerick, Ireland

Received: 21 August 2020; Accepted: 29 August 2020; Published: 1 September 2020

Over the past two decades, considerable research efforts in academia and industry have gone into pharmaceutical cocrystals [1,2]. As a result, a large number of studies on both fundamental aspects and applications of cocrystallisation have been published, and a few cocrystals are now on the market or in clinical trial phases, e.g., sacubitril-disodium valsartan-water (Entresto^{TM}), escitalopram oxalate-oxalic acid (Lexapro$^{®}$), ertuglifozin-L-pyroglutamic acid and tramadol-celecoxib. The FDA defines pharmaceutical cocrystals as "crystalline materials composed of two or more different molecules, typically active pharmaceutical ingredient (API) and cocrystal formers ('coformers'), in the same crystal lattice" [3]. Cocrystallisation is an attractive strategy to modify and improve the physicochemical properties of an API without making covalent changes to the drug molecule itself. Very often cocrystals are designed to tackle the poor dissolution behaviour and low bioavailability of Biopharmaceutics Classification System (BCS) class II and IV drugs that make up 70% of the drug candidates in the development pipeline [4]. However, chemical stability, hygroscopicity, mechanical, and flow properties have also been improved through cocrystal formation [1,2]. Furthermore, cocrystallisation can be used as a purification and enantiomeric separation method [5].

This Special Issue includes eight original research articles that highlight the relevance and versatility of pharmaceutical cocrystals in drug delivery.

Machado Cruz et al. report a new cocrystal of the poorly water-soluble antifungal agent itraconazole [6]. They carried out a comprehensive study of the solid-state properties and the formation of the itraconazole-terephthalic acid cocrystal. The cocrystal is stable in aqueous solution and comparison with previously described itraconazole cocrystals showed a correlation of the intrinsic and powder dissolution rates with the solubility of the coformer. The dissolution behaviour of physical mixtures of the cocrystal and common excipients was also analysed.

Nugrahani et al. prepared the mono- and tetrahydrate of the salt cocrystal diclofenac sodium-L-proline [7]. The hydrates were characterised by single crystal X-ray analysis and were shown to have higher solubilities and dissolution rates than the sodium salt of diclofenac acid and the anhydrous diclofenac acid-L-proline cocrystal. The release of water on drying led to the dissociation of the salt cocrystal into a physical mixture of diclofenac acid and L-proline. Interestingly, this process was reversible. When the dried sample was kept at 72% relative humidity and 25 °C, diclofenac sodium-L-proline tetrahydrate was restored.

A paper by Buol et al. describes the first cocrystals of the nootropic drug nefiracetam [8]. A large cocrystal screen with 133 different coformers using liquid-assisted grinding led to the identification of 13 new cocrystals that were characterised by single-crystal X-ray diffraction. The study illustrates how solid-state properties—in this case, the melting point—can be varied over a wide range by changing the coformer. Three cocrystals with biocompatible coformers were subjected to a more comprehensive screen including solvent evaporation, slurrying and cooling crystallisation. The discovery of additional solid-state forms demonstrates the importance of a thorough screening for multiple cocrystal forms. The solubilities and dissolution properties of the new cocrystals were investigated.

Kale et al. studied the effect of cocrystallisation on the tabletability of rivaroxaban and found an improved tabletability for rivaoxaban-malonic acid [9]. The tabletability order malonic acid <

rivaboxaban < rivaboxaban-malonic acid could be rationalised with the crystal packing, specifically the absence or presence of slip planes, slip plan topology, the degree of intermolecular interactions and d-spacing. Rivaroxaban contains slip planes with a flat-layered topology and with a zigzag layer topology. The crystal structure-mechanical property relationships found in this study shine light on the way crystals that contain more than one slip-plane system deform.

Wroblewska et al. introduced the new coformer 1-hydroxy-4,5-dimethyl-imidazole 3 oxide [10]. They used high resolution-solid-state NMR to investigate the cocrystal formation with barbituric and thiobarbituric acid during ball-milling. The structures of the new coformer and cocrystals were studied by ^{13}C CP/MAS, ^{15}N CP/MAS and ^{1}H Very Fast MAS NMR in combination with single-crystal X-ray analysis. The effect of the polymorphic and tautomeric form of barbituric/thiobarbituric acid on the cocrystallisation was evaluated. The cocrystals showed no cytotoxicity in HeLa and 293T cells at concentrations of up to 100 µM indicating good biocompatibility of 1-hydroxy-4,5-dimethyl-imidazole 3 oxide. However, the coformer failed to give a significant increase in solubility.

Salas-Zuniga et al. studied the effect of hydroxypropyl methylcellulose and methylcellulose on the dissolution behaviour of two nitazoxanide cocrystals [11]. Using polymer-based powder formulations of nitazoxanide-succinic acid, they achieved a significant improvement of the apparent solubility of nitazoxanide compared to formulations of the pure API. It was suggested that the solubility enhancement was due to a polymer-induced delay of nucleation and crystal growth.

Witika et al. report a cocrystal of lamivudine and zidovudine [12]. The dual-drug cocrystal was characterised by X-ray powder diffraction, Raman spectroscopy, FT-IR spectroscopy, differential scanning calorimetry and energy-dispersive X-ray spectroscopy. Surfactants were applied to produce and stabilise nano-cocrystals with specific, pre-defined critical quality attributes such as particle size, polydispersity index, and zeta potential to exploit the advantages of nano-sized drug delivery systems. In a follow-up paper, the same authors describe a Design of Experiment approach to optimise the cold sonochemical synthesis of the lamivudine-zidovudine nano-cocrystals in the presence of surfactants and polymers [13]. The nano-cocrystals proved to be less cytotoxic in HeLa cells than a physical mixture of the two APIs.

Funding: This research received no external funding.

Conflicts of Interest: The author declares no conflict of interest.

References

1. Karagianni, A.; Malamatari, M.; Kachrimanis, K. Pharmaceutical Cocrystals: New Solid Phase Modification Approaches for the Formulation of APIs. *Pharmaceutics* **2018**, *10*, 18. [CrossRef] [PubMed]
2. Karimi-Jafari, M.; Padrela, L.; Walker, G.M.; Croker, D.M. Creating Cocrystals: A Review of Pharmaceutical Cocrystal Preparation Routes and Applications. *Cryst. Growth Des.* **2018**, *18*, 6370–6387. [CrossRef]
3. Food and Drug Administration. Regulatory Classification of Pharmaceutical Co-Crystals, Guidance for Industry. February 2018. Available online: https://www.fda.gov/media/81824/download (accessed on 20 August 2018).
4. Williams, H.D.; Trevaskis, N.L.; Charman, S.A.; Shanker, R.M.; Charman, W.N.; Pouton, C.W.; Porter, C.J.H. Strategies to address low drug solubility in discovery and development. *Pharmacol. Rev.* **2013**, *65*, 315–499. [CrossRef] [PubMed]
5. Springuel, G.; Leyssens, T. Innovative Chiral Resolution Using Enantiospecific Co-Crystallization in Solution. *Cryst. Growth Des.* **2012**, *127*, 3374–3378. [CrossRef]
6. Machado Cruz, R.; Boleslavská, T.; Beránek, J.; Tieger, E.; Twamley, B.; Santos-Martinez, M.J.; Dammer, O.; Tajber, L. Identification and Pharmaceutical Characterization of a New Itraconazole Terephthalic Acid Cocrystal. *Pharmaceutics* **2020**, *12*, 741. [CrossRef]
7. Nugrahani, I.; Kumalasari, R.A.; Auli, W.N.; Horikawa, A.; Uekusa, H. Salt Cocrystal of Diclofenac Sodium-L-Proline: Structural, Pseudopolymorphism, and Pharmaceutics Performance Study. *Pharmaceutics* **2020**, *12*, 690. [CrossRef] [PubMed]

8. Buol, X.; Robeyns, K.; Caro Garrido, C.; Tumanov, N.; Collard, L.; Wouters, J.; Leyssens, T. Improving Nefiracetam Dissolution and Solubility Behavior Using a Cocrystallization Approach. *Pharmaceutics* **2020**, *12*, 653. [CrossRef] [PubMed]

9. Kale, D.P.; Puri, V.; Kumar, A.; Kumar, N.; Bansal, A.K. The Role of Cocrystallization-Mediated Altered Crystallographic Properties on the Tabletability of Rivaroxaban and Malonic Acid. *Pharmaceutics* **2020**, *12*, 546. [CrossRef] [PubMed]

10. Wróblewska, A.; Śniechowska, J.; Kaźmierski, S.; Wielgus, E.; Bujacz, G.D.; Mlostoń, G.; Chworos, A.; Suwara, J.; Potrzebowski, M.J. Application of 1-Hydroxy-4,5-Dimethyl-Imidazole 3-Oxide as Coformer in Formation of Pharmaceutical Cocrystals. *Pharmaceutics* **2020**, *12*, 359. [CrossRef] [PubMed]

11. Salas-Zúñiga, R.; Rodríguez-Ruiz, C.; Höpfl, H.; Morales-Rojas, H.; Sánchez-Guadarrama, O.; Rodríguez-Cuamatzi, P.; Herrera-Ruiz, D. Dissolution Advantage of Nitazoxanide Cocrystals in the Presence of Cellulosic Polymers. *Pharmaceutics* **2020**, *12*, 23. [CrossRef] [PubMed]

12. Witika, B.A.; Smith, V.J.; Walker, R.B. A Comparative Study of the Effect of Different Stabilizers on the Critical Quality Attributes of Self-Assembling Nano Co-Crystals. *Pharmaceutics* **2020**, *12*, 182. [CrossRef] [PubMed]

13. Witika, B.A.; Smith, V.J.; Walker, R.B. Quality by Design Optimization of Cold Sonochemical Synthesis of Zidovudine-Lamivudine Nanosuspensions. *Pharmaceutics* **2020**, *12*, 367. [CrossRef] [PubMed]

 *pharmaceutics*

Article

Identification and Pharmaceutical Characterization of a New Itraconazole Terephthalic Acid Cocrystal

Ricardo Machado Cruz [1], **Tereza Boleslavská** [2,3], **Josef Beránek** [2], **Eszter Tieger** [2], **Brendan Twamley** [4], **Maria Jose Santos-Martinez** [1,5], **Ondřej Dammer** [2] and **Lidia Tajber** [1,*]

[1] School of Pharmacy and Pharmaceutical Sciences, Trinity College Dublin, Dublin 2, Ireland; cruzr@tcd.ie (R.M.C.); santosmm@tcd.ie (M.J.S.-M.)

[2] Zentiva, k.s., U Kabelovny 130, 102 37 Prague, Czech Republic; tereza.boleslavska@zentiva.com (T.B.); josef.beranek@zentiva.com (J.B.); tiegereszter@gmail.com (E.T.); ondrej.dammer@zentiva.com (O.D.)

[3] Department of Chemical Engineering, University of Chemistry and Technology, Prague, Technická 5, 166 28 Prague, Czech Republic

[4] School of Medicine, Trinity College Dublin, Dublin 2, Ireland; twamleyb@tcd.ie

[5] School of Chemistry, Trinity College Dublin, Dublin 2, Ireland

* Correspondence: ltajber@tcd.ie; Tel.: +353-1-896-2787

Received: 17 July 2020; Accepted: 31 July 2020; Published: 6 August 2020

Abstract: The crystallization of poorly soluble drug molecules with an excipient into new solid phases called cocrystals has gained a considerable popularity in the pharmaceutical field. In this work, the cocrystal approach was explored for a very poorly water soluble antifungal active, itraconazole (ITR), which was, for the first time, successfully converted into this multicomponent solid using an aromatic coformer, terephthalic acid (TER). The new cocrystal was characterized in terms of its solid-state and structural properties, and a panel of pharmaceutical tests including wettability and dissolution were performed. Evidence of the cocrystal formation was obtained from liquid-assisted grinding, but not neat grinding. An efficient method of the ITR–TER cocrystal formation was ball milling. The stoichiometry of the ITR–TER phase was 2:1 and the structure was stabilized by H-bonds. When comparing ITR–TER with other cocrystals, the intrinsic dissolution rates and powder dissolution profiles correlated with the aqueous solubility of the coformers. The rank order of the dissolution rates of the active pharmaceutical ingredient (API) from the cocrystals was ITR–oxalic acid > ITR–succinic acid > ITR–TER. Additionally, the ITR–TER cocrystal was stable in aqueous conditions and did not transform to the parent drug. In summary, this work presents another cocrystal of ITR that might be of use in pharmaceutical formulations.

Keywords: itraconazole; terephthalic acid; cocrystal; crystal structure; mechanochemistry; solid-state; thermal analysis; wettability; dissolution

---

1. Introduction

A pharmaceutical cocrystal can be defined as a multicomponent crystal wherein at least one component is the active pharmaceutical ingredient (API) and is in a well-defined stoichiometric ratio and bonded by non-covalent interactions with the other component(s), i.e., the coformer(s) [1,2]. These interactions are mainly intermolecular hydrogen bonds between functional groups forming supramolecular synthons, and moreover, additional weaker interactions such as van der Waals forces, π-stacking or halogen bonds may also help to stabilize the cocrystal structure [3,4]. The purpose of synthesizing pharmaceutical cocrystals is to improve the characteristics of APIs, such as the aqueous solubility, dissolution rates and stability [5].

Itraconazole (ITR) is an API with a broad-spectrum antifungal activity used for the treatment of topical and systemic mycoses as well as for prophylaxis in immunosuppressed patients [6,7]. ITR, due to a low aqueous solubility but a high systemic absorption, is considered a class II drug according to the biopharmaceutics classification system (BCS) [8]. Indeed, ITR has a very low solubility in water (around 1 ng/mL), which increases in an acidic medium (4 µg/mL) [9], thus the poor solubility is the limiting factor for ITR absorption. To improve the poor physicochemical properties, ITR can be converted into disordered forms such as liquid crystalline [10–12] and amorphous structures, as well as polymer-based solid dispersions [12]. The commercial solid dosage formulation of ITR, Sporanox®, is available as pellets enclosed in oral capsules [13]. The pellets are manufactured by spray-layering a solution of ITR and hydroxypropyl methylcellulose (HPMC) on sucrose beads. Then, the drug-coated beads are treated with a seal coating polymer layer (polyethylene glycol (PEG) 20,000 Da) to prevent the sticking of the beads. The API in these pellets is in the amorphous state [13]. Since the physical stability of drug in an amorphous phase can be short-lived, a crystalline, but a better soluble form of ITR would be preferred.

The development of ITR cocrystals was first achieved combining this API with aliphatic dicarboxylic acids used as coformers, including fumaric acid, succinic acid, L-malic acid, L-tartaric acid, D-tartaric acid and D,L-tartaric acid [14]. Remenar's work reported a remarkable improvement of the ITR dissolution with an approximately 20-fold enhancement of the cocrystals prepared with L-malic and L-tartaric acids, which had dissolution profiles comparable to the commercial form of ITR, Sporanox® capsules. In another work, Shevchenko et al. [15] produced ITR cocrystals with coformers of different carbon chain lengths and determined that linear aliphatic dicarboxylic acids should have no more than seven carbon atoms in their chains to successfully form a cocrystal with ITR. That work also demonstrated that cocrystallisation is an efficient approach to improve the dissolution rates of ITR. However, a recent study presented that an ITR cocrystal with an aliphatic dicarboxylic acid comprising more than seven carbon atoms, suberic acid, can be prepared by rapid solvent evaporation and spray drying, resulting in the material being able to dissolve rapidly. This cocrystal had a 39 times faster intrinsic dissolution rate than the crystalline ITR [16].

Little is published on the structural analysis of ITR cocrystals, nevertheless, such studies on the ITR and succinic acid (SUC) cocrystal revealed that this form has a trimeric building block, where two molecules of the API are oriented antiparallel to one another and bridged by the coformer. The main interactions found to be responsible for the intermolecular arrangement of this structure were H-bonds formed by the hydroxyl moiety located on both sides of the acid and one of the nitrogen atoms of the 1,2,4-triazole ring of each ITR molecule (Figure 1) [14].

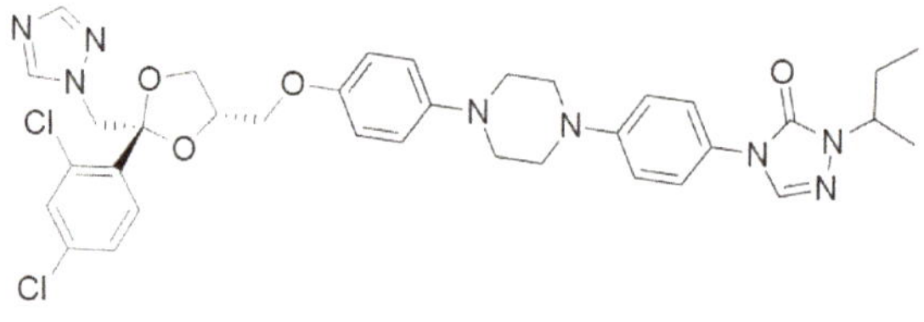

(a)

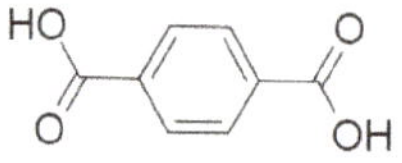

(b)

Figure 1. Molecular structures of (**a**) itraconazole (ITR) and (**b**) terephthalic acid (TER).

Terephthalic acid (TER) is a benzenedicarboxylic acid with the carboxylic groups attached in positions one and four (Figure 1). TER has very low toxicity and has been employed in the cocrystallization of a few APIs, such as apovincamine [17], betulin [18], isoniazid [19] and gabapentin [20]. Considering that most investigations on the identification of ITR cocrystals are limited to aliphatic dicarboxylic acids [14–16,21], coformers with other architecture appears to be unexplored. Therefore, this work carried out an experimental screening aiming to identify a possible new cocrystal of ITR combining this API with TER to ascertain the relevant molecular elements in the coformers that enable the formation of new cocrystal phases of ITR. This new cocrystalline form was extensively evaluated regarding its solid-state characteristics, as was the impact ITR cocrystallisation had on a range of pharmaceutical properties. Finally, the cocrystal dissolution was studied including an intrinsic dissolution study, a simple mixture with lactose as well as a mixture comprising excipients used in the commercial ITR formulation, comparing the performance of the new cocrystal with those based on aliphatic dicarboxylic acids.

2. Materials and Methods

2.1. Materials

Itraconazole (ITR) was purchased from Glentham Life Sciences Ltd. (Whitshire, UK). MeOH (HPLC grade) and terephthalic acid (TER) were purchased from Sigma-Aldrich (Arklow, Ireland). All other ingredients, such as solvents, solution and buffer components, as well as polymers, were kindly provided by Zentiva (Prague, Czech Republic).

2.2. Methods

2.2.1. Neat Grinding (NG) and Liquid-Assisted Grinding (LG)

For these preparations, 40 mg of ITR and an amount of TER corresponding to 2:1, 1:1 and 1:2 of API-coformer molar ratios were carefully weighted. Afterwards, the compounds were ground for 30 s using an agate mortar and pestle.

For liquid-assisted grinding, two drops of methanol were added to the mixture of powders.

2.2.2. Cocrystallisation by Slurrying

In this method, solutions of ITR at 0.74 and 2.20 mg/mL were prepared in methanol (MeOH) and acetone, respectively. Then, a 1 mL aliquot of each solution was transferred to a 1.5 mL glass vial and a quantity of TER was added in a 1:50 API-coformer molar ratio. Then, the hermetically closed vials containing the suspensions were mixed at 50 °C for 8 h and thereafter for next 4 days at room temperature using an IKA RT 15 magnetic stirrer (Germany). Afterwards, the slurries were dried at 30 °C and 100 mbar in a 3608-6CE vacuum oven (ThermoFisher Scientific™, Waltham, MA, USA).

2.2.3. Cocrystallisation by Ball Milling (BM)

A quantity of 300 mg of ITR was weighted and added to an amount of each coformer (oxalic acid, succinic acid and TER) corresponding to a 2:1 API-coformer molar ratio. Then, the powders were transferred to a 25 mL stainless-steel grind jar containing two 15 mm stainless-steel balls. Before grinding, two drops of acetone were added to the powders. The mixtures were ground in two cycles of 10 min at 25 Hz using a Retsch Mixer Mill MM 200 (Haan, Germany).

2.2.4. Cocrystallisation by Slow Evaporation

A solution of ITR and TER was prepared by weighing 5.96 mg of the API and 0.70 mg of the coformer to prepare a 2:1 mole/mole mixture of the components. The powders were transferred into a glass vial and solubilized in 10 mL of methanol by sonication in a U300 H ultrasonic bath (Ultrawave, Rumney, UK) to obtain a clear solution. Then, the solution was filtered using a 0.45 μm PTFE syringe filter (Fisher Scientific, Loughborough, UK). Parafilm was used to cover the top of the vial and small holes were pierced to allow for solvent evaporation. The solution was left at room conditions until crystallization occurred.

2.2.5. Freeze Drying of Itraconazole (ITR)

Firstly, a 10 mg/mL solution of ITR in dioxane was prepared by weighing 1 g of the API and solubilizing it in 100 mL of the solvent. The solution was divided into three round-bottomed flasks and frozen using liquid nitrogen while rotating using a Rotavapor R-205 (Büchi, Flawil, Switzerland). The samples were dried for 18 h using a freeze drier, ALPHA 2-4 LSC (Martin Christ, Osterode am Harz, Germany), with manifolds for a connection of NS 29/32 flasks under a vacuum of 2×10^{-3} mbar, between 5 and 6 m^3/h suction and ice condenser adjusted to -85 °C. No secondary drying step was applied.

2.2.6. Differential Scanning Calorimetry (DSC)

A thermal analysis was performed by carefully weighting dried samples and placing them in 40 µL aluminum pans that were sealed with a lid containing three vent holes. The samples were subjected to DSC runs in a temperature ranging from 25 to 400 °C with a heating rate of 10 °C/min using a Mettler Toledo DSC 822 e/700 (Greifensee, Switzerland) under nitrogen purge [22]. An empty aluminum pan was used as a reference. The equipment was calibrated with an indium standard.

2.2.7. Powder X-ray Diffraction (PXRD)

PXRD patterns were obtained with a laboratory X'PERT PRO MPD (PANalytical, Almelo, Netherlands) diffractometer with CuKα (λ= 1.542 Å) radiation. The generator was operated at an excitation voltage of 45 kV and anodic current of 40 mA. The following scan parameters were utilized: scan type—gonio, measurement range of 2–40° 2θ, step size of 0.02° 2θ and the time per step was 200 s. The samples were placed on a zero-background silica sample holder. Alternatively, PXRD measurements were performed using a Rigaku Miniflex II, desktop X-ray diffractometer (Tokyo, Japan) equipped with a CuKα (λ = 1.54 Å) radiation X-ray source. Dried samples were mounted on a low-background silicon sample holder and scanned over a 2θ range of 2–40 degrees [23].

2.2.8. Single Crystal X-ray Analysis

A monocrystal of ITR–TER with approximate dimensions of 0.030 mm × 0.140 mm × 0.150 mm was used for the X-ray crystallographic analysis. The X-ray intensity data were measured at 100 ± 2 K on a Bruker Apex Kappa Duo (Billerica, MA, USA) with an Oxford Cobra Cryosystem low-temperature device (Oxford, UK) using a MiTeGen micromount (Ithaca, NY, USA). Bruker APEX software was used to correct for Lorentz and polarization effects. The crystallographic data were analyzed using Mercury 2020.1 and CrystalExplorer (ver. 17.5) [24] software.

2.2.9. Fourier-Transform Infrared Spectroscopy (FTIR) and Raman Spectroscopy

The dried powders were subjected to FT–IR spectroscopy on a PerkinElmer Spectrum 100 (Waltham, MA, USA), equipped with a universal attenuated total reflection (ATR) device and a ZnSe crystal. The FT–IR spectra of the samples were recorded in a wavelength range from 500 to 4000 cm^{-1}. The spectra were acquired by averaging 10 scans taken with a resolution of 4 cm^{-1}.

The Raman spectra of powders were measured directly in glass vials using a Raman Spectrometer RFS 100/S (Bruker, Billerica, MA, USA). The spectra were acquired by averaging 64 scans taken with a resolution of 4 cm^{-1} and laser power of 250 mW.

2.2.10. Morphological Analysis

A Zeiss Supra variable Pressure Field Emission Scanning Electron Microscope (SEM, Ulm, Germany) equipped with a secondary electron detector and an accelerating voltage of 5 kV was used for the morphological examination. The produced powder samples were placed on carbon tabs attached to aluminum stubs and sputter coated with gold/palladium under vacuum before analysis [10].

2.2.11. Contact Angle Measurements

Powdered samples were compacted into 4.5 cm in diameter disks and placed on the lifting table of a Drop Shape Analyzer DSA 25 (Krüss, Hamburg, Germany). Then, an automated dosing syringe containing water at 20 °C was used to deposit a single drop of 14 μL on the surface of the disks. The images of the water drop on the surface of the disks were recorded for 10 min by a high-resolution camera and processed by ADVANCE software ver. 1.9 (Krüss) to calculate the contact angle. Each sample had the contact angle measured in duplicate.

2.2.12. Intrinsic Dissolution Rate (IDR) Study

Disks 8 mm in a diameter were prepared by compressing 50 ± 2 mg of powder in a stainless-steel cylindrical die system at approximately 100 kg/cm^2 for 120 s. Then, the opposite side of the steel die was sealed using a rubber plug, leaving a 0.503 cm^2 surface of the disk exposed. The stainless-steel dies were used as the disk holders and were loaded automatically by the robotic arm of the Pion inForm (Pion, Forest Row, UK) and immersed in the dissolution media. For each test, 40 mL of the medium (acetate-phosphate buffer comprising 150 mM NaCl) was preheated to 37 °C and the pH was adjusted with 0.5 M HCl to 1.2. The agitation was set to 100 rpm. A spectral scan (190–720 nm) was collected every 30 s and the concentration was calculated against the calibration curve obtained previously under identical conditions.

2.2.13. Dissolution Analysis

2.2.13.1. Powder Dissolution of ITR Systems Mixed with Lactose

For this procedure, samples were prepared by weighing an amount of ITR (starting material), freeze dried ITR (FD ITR), ITR–TER, itraconazole-oxalic acid cocrystals (ITR–OXA) and itraconazole-succinic acid cocrystals (ITR–SUC) containing an equivalent of 100 mg of ITR and mixing with lactose monohydrate in a 1:6 *w/w* API:excipient ratio to allow wettability and dispersibility in the liquid medium. Sporanox$^®$ was used as the reference formulation and the pellets were removed from the capsule before the use in the dissolution experiments. The dissolution analysis was performed using a solution prepared by mixing 33 mM NaCl with 67 mM HCl (artificial gastric juice (AGJ)) containing 0.05% (*v/v*) of Tween 20. The pH of this mixture was then adjusted to 1.2 with 0.5 M HCl. The experiments were carried out using a standard USP II dissolution apparatus (Sotax, Aesch, Switzerland) attached to a UV-Vis spectrophotometer Specord 200 Plus (Analytik Jena, Jena, Germany). The powders were added directly to the media (900 mL) kept at 37 °C and agitated at 75 rpm for the first 45 min and at 150 rpm for the final 15 min. The aliquots were automatically taken at predefined time points (2, 5, 10, 15, 20, 25, 30, 45, 50 and 60 min) and the ITR concentration was assessed by measuring the absorbance at 255 nm.

2.2.13.2. Powder Dissolution of ITR Systems Mixed with Other Excipients

A second dissolution test evaluated the dissolution of the API when physically mixed with the same excipients as those present in Sporanox$^®$ (Jassen, Beerse, Belgium). For this purpose, 350 mg of ITR (starting material) and FD ITR or the amount of the ITR–TER, ITR–OXA and ITR–SUC cocrystals containing the equivalent of 350 mg of the API were carefully weighted and mixed with the other excipients in the concentrations listed in Table 1 for 5 min in 200 mL plastic bottles using a Turbula$^®$ mixer (WAB group, Muttenz, Switzerland).

The dissolution analysis was carried out as described in Section 2.2.13.1, using 900 mL peak vessels in a dissolution apparatus 708-DS (Agilent, Lexington, MA, USA) coupled to an Agilent UV-Vis spectrophotometer Cary 60.

Table 1. Composition of the powders used in the dissolution study.

Excipient (%, *w/w*)	Formulation				
	ITR	FD ITR	ITR–OXA	ITR–SUC	ITR–TER
API	21.74	21.74			
Cocrystal			23.22	23.13	23.70
Sucrose	41.74	41.74	40.95	41.0	40.69
HPMC [(*)]	32.61	32.61	31.99	32.03	31.79
PEG [(**)]	3.91	3.91	3.84	3.84	3.82
Total (%)	100	100	100	100	100
Powder weight (mg) [(***)]	460	460	501.1	498.32	514.2

(*) Hydroxypropyl methyl cellulose 2910 (5 mPa.s (HPMC)); (**) polyethylene glycol 6000 Da (PEG); (***) mass of the powder containing 100 mg of ITR used in the dissolution test.

2.2.14. Statistical Analysis

The statistical analysis of the data was carried out using GraphPad Prism® for Windows, version 5.01, applying a Student's *t*-test or one-way ANOVA with Tukey's post-test with 95% of confidence when appropriate. Statistical significance was when $p < 0.05$.

3. Results and Discussion

3.1. Characterization of ITR and Terephthalic Acid (TER) Mixtures Following Neat and Liquid-Assisted Grinding

Mechanochemical methods of cocrystal production have gained a considerable interest in recent times [25,26]. Therefore, as the first approach, a neat, solvent-free (NG) grinding of ITR and TER in a few stochiometric ratios, followed by the liquid-assisted grinding (LG) was performed. The ITR "as received" material was identified as form I of itraconazole [27]. The PXRD analysis of the samples prepared by neat grinding (NG (Figure 2)) revealed that the mixtures post-processing had similar diffractograms to those of ITR with two Bragg peaks at 17.5 and 17.95° 2θ, corresponding to TER. The samples prepared by LG had additional, but weak, Bragg peaks at 3.5, 7.0 and 21.2° 2θ, which were absent in the parent materials, indicating the formation of a new phase. Interestingly, the TER component in all samples post NG and LG appeared to be at least partially amorphous, as evidenced by the weak or absent diffraction peaks of TER. Therefore, the LG method is more efficient in producing the ITR–TER cocrystal than NG most likely because the cocrystallisation process by LG was facilitated by the presence of methanol. Indeed, the use of solvents in cocrystallisation screening is common and the impact of the solvent is described as catalytic, since it is used in a very small quantity and is also not part of the final cocrystal [28].

A DSC analysis of ITR and TER on their own showed that the drug melted at around 166 °C [12], while TER melted with sublimation at around 350 °C [29] (Figure 3). A thermal analysis of the binary ITR–TER mixtures post-processing showed thermograms with a sharp endothermic peak at 198 °C (Figure 3), which was absent in the parent compounds and indicated the melting of the new phase. However, all these mixtures also had an endothermic peak with an onset at 164 °C (Figure 3), assigned to the melting of ITR. This peak in the ITR–TER 1:1 (LG) and ITR–TER 1:2 (LG) systems was almost imperceptible, with enthalpies of 2.6 and 0.7 J/g, respectively. In contrast, the ITR peak in the samples ITR–TER 1-1 (NG) and ITR–TER 1-2 (NG) had an enthalpy of 13.5 and 18.4 J/g, respectively, indicating that the methanol used in LG contributed to a greater conversion of ITR into the new phase. A small exothermic peak was detected for all NG samples and ITR–TER 1:2 (LG) immediately followed the ITR peak, consistent with the previous studies on the thermal behavior of binary mixtures where the formation of a cocrystal was observed upon heating [30]. Thus, in relation to NG mixtures, the cocrystal melting peak, visible in DSC traces, is of the new form which appears upon heating. Additionally, what can be concluded from the DSC study is that a broad endothermic peak with an onset at around 310 °C for the ITR–TER 1:1 samples and an onset at around 320 °C for the ITR–TER 1:2 systems was detected, assigned to the excess of TER. It may suggest that the potential stoichiometry of the ITR–TER cocrystal might be 2:1.

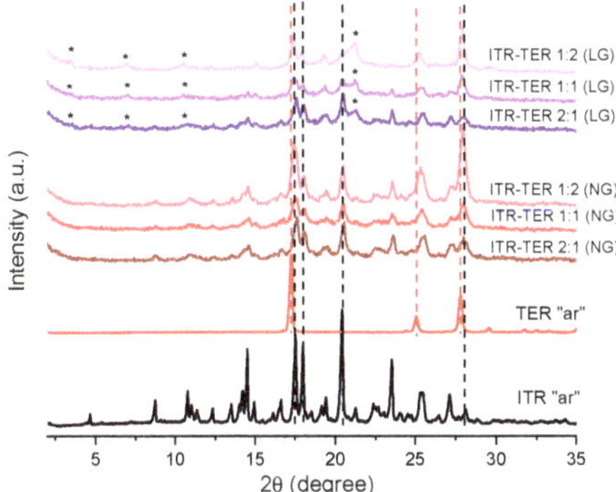

Figure 2. Powder X-ray diffraction (PXRD) of the ITR and TER starting material as the received powders ("ar") and binary ITR and TER mixtures following neat grinding (NG) and liquid-assisted grinding (LG). * Indicates the distinct peak of the new phase (cocrystal), broken black lines show the position of the ITR diffraction peaks, while the broken red lines show the position of the TER diffraction peaks.

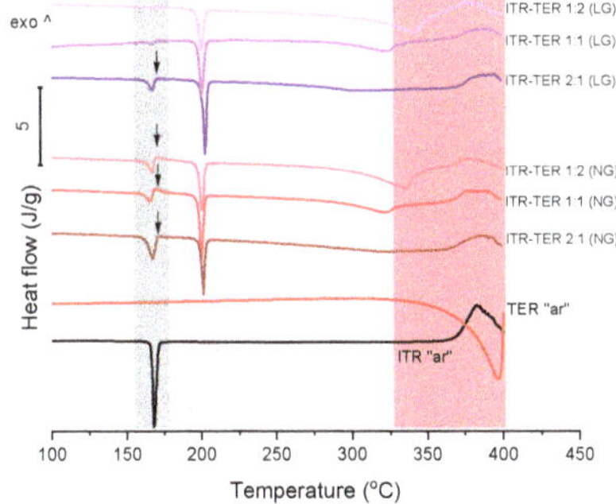

Figure 3. Differential scanning calorimetry (DSC) of the ITR and TER starting material, as received powders ("ar") and binary ITR and TER mixtures following neat grinding (NG) and liquid-assisted grinding (LG). The area highlighted in grey indicates the ITR melting region, in purple the cocrystal melting range and in light red the TER melting and degradation. Arrows indicate the presence of an exothermic peak immediately following the ITR melting event.

3.2. Properties of ITR and TER Samples Made by Slurring, Evaporation and Ball Milling Methods

A PXRD analysis of samples obtained by the slurring of ITR and TER in methanol and acetone was very similar to that of TER and indicated an incomplete conversion to the cocrystal, with only weak intensity Bragg peaks of the new phase visible at approximately 7.0, 10.4, 12.4° 2θ (Figure 4). A more successful method was the slow evaporation of ITR and TER from methanol, and the diffraction pattern of the 2:1 ITR–TER system is presented in Figure 4. A number of diffraction peaks characteristic of the cocrystal phase can be discerned, however the peak at 15.0° 2θ can be described as the ITR starting material powder. The method that gave the purest cocrystal, based on the PXRD results, was ball milling (Figure 4). The ITR–TER 2:1 system had a diffraction pattern distinct from those of the starting material powders, particularly due to the peaks at 3.5, 7.0, 10.5, 12.4, 17.8, 19.3 and 21.2° 2θ. Therefore, ball milling, as the most efficient method of cocrystal production, was used to prepare the cocrystal for further characterizations.

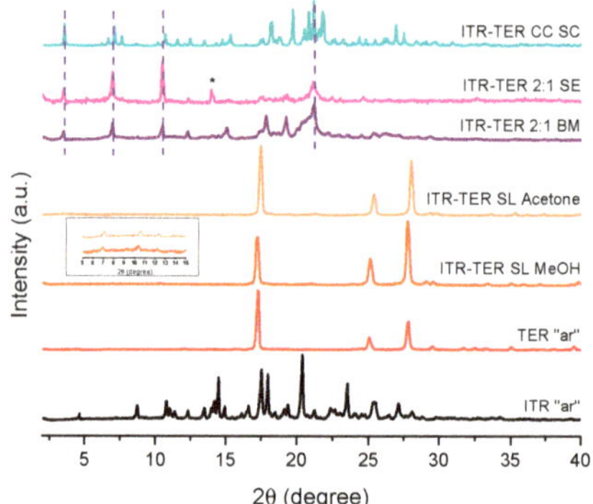

Figure 4. PXRD patterns of the ITR and TER starting material, as received powders ("ar"), samples produced by the slurry method from methanol (ITR–TER SL MeOH) and acetone (ITR–TER SL acetone), the ball milled sample containing ITR and TER in a 2:1 stoichiometric ratio (ITR–TER 2:1 BM), the sample obtained by the slow evaporation of ITR and TER in a 2:1 stoichiometric ratio from methanol (ITR–TER 2:1 SE) and a powder diffraction pattern of the ITR–TER cocrystal calculated from the single crystal X-ray data (ITR–TER CC SC). * Indicates the peak of ITR, broken lines indicate the position of key diffraction peaks of the cocrystal, while the insert shows the presence of the cocrystal peaks in the samples obtained by the slurry method.

A DSC analysis (Figure 5) confirmed the above XPRD data. The samples prepared by the slurring of ITR and TER in methanol and acetone had low magnitude peaks with an onset at 195 °C, which corresponded to the melting of the cocrystal phase. In both samples this peak had the normalized enthalpy of transition of 8.1 J/g. DSC also confirmed that these samples were mainly composed of TER. The sample obtained by solvent evaporation displayed a characteristic event of cocrystal melting, although the peak was broader than that of the ball milled sample. The thermal analysis of the ITR–TER 2:1 ball milled system (Figure 5) verified that ball milling was more efficient in producing the cocrystal, with a sharp melting peak with a normalized enthalpy of 83.7 J/g.

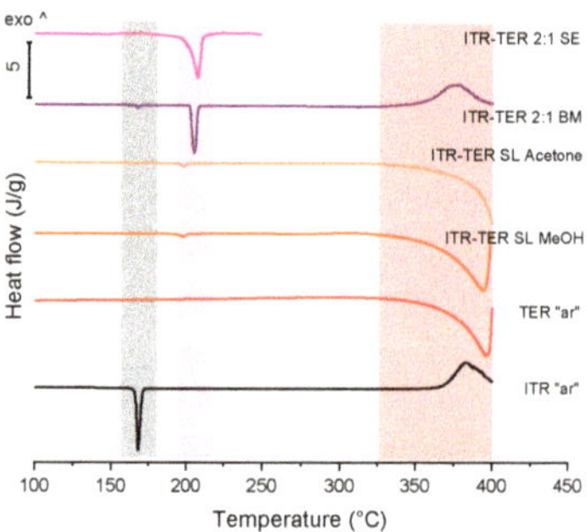

Figure 5. DSC thermograms of the ITR and TER starting material, as received powders ("ar"), samples produced by slurry method from methanol (ITR–TER SL MeOH) and acetone (ITR–TER SL acetone), the ball milled sample containing ITR and TER in the 2:1 stoichiometric ratio (ITR–TER 2:1 BM) and the sample obtained by slow evaporation of ITR and TER in the 2:1 stoichiometric ratio from methanol (ITR–TER 2:1 SE). The area highlighted in grey indicates the ITR melting region, in purple the cocrystal melting range and in light red the TER melting and degradation.

3.3. Infrared and Raman Spectroscopy

The IR spectrum of the ITR–TER cocrystal showed an intense band at 1705 cm^{-1}, which was also present in the ITR and TER spectra, at 1698 cm^{-1} and 1675 cm^{-1}, respectively (Figure 6a). Strong peaks in this range are normally caused by the stretching of C=O [31]. As this peak appears at a higher wavenumber in the cocrystal than in the parent compounds, this suggests that the new H-bond in the cocrystal might be weaker than the one in TER, as ITR does not form any intermolecular H-bonds. Another area of interest was the fingerprint region, between 1200 and 1300 cm^{-1}. In the TER spectrum, a broad strong peak at 1281 cm^{-1}, most likely from the –C–O–H vibration, was seen, while in ITR, a peak at 1271 cm^{-1} was assigned to aromatic C–N stretching [32,33]. In the cocrystal, a number of medium intensity bands were detected around this area.

Raman spectroscopy confirmed the presence of the new phase and the spectrum of the cocrystal (Figure 6b) was significantly different when compared to those of ITR and TER, which indicate interactions between the API and the coformer forming supramolecular interactions [34]. In the pure coformer, H-bonded dimers are formed by the –COOH groups, with the C=O moiety acting as the H-bond acceptor and the –OH moiety being the H-bond donor. In the cocrystal, the acid dimer arrangement is disrupted, which results in a shift of the peak assigned to the C=O stretching band to 1696 cm^{-1} in the cocrystal spectrum, while in the conformer, the band is located at 1633 cm^{-1}. This effect was also observed by Shevchenko et al. [15], who investigated the formation of ITR cocrystals with aliphatic dicarboxylic acid of varying chain lengths, including succinic acid, malonic acid, and oxalic acid. In that work, the authors detected a peak shift of the C=O vibration in all produced cocrystals.

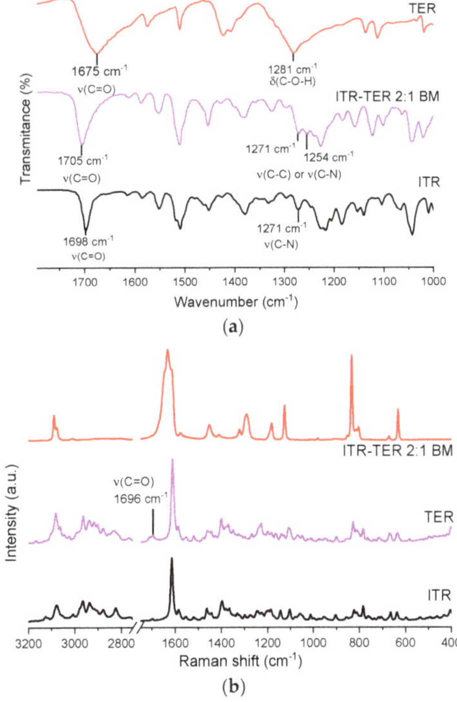

Figure 6. (a) Fourier-transform infrared spectroscopy (FT-IR) spectra of the ITR and TER starting material powders as well as the ITR–TER cocrystal prepared by ball milling (ITR–TER 2:1 BM); (b) Raman spectra of the ITR and TER starting material powders as well as the ITR–TER cocrystal prepared by ball milling (ITR–TER 2:1 BM). ν–stretching.

3.4. Crystallographic Analysis

Single crystal X-ray diffraction (SC-XRD) showed that the cocrystal had a triclinic unit cell composed of four molecules (Z = 4) in a P$\bar{1}$ space group (Table S1). The asymmetric unit crystal was formed by two ITR molecules and two halves of a TER molecule. The stoichiometry of the cocrystal was determined as 2 moles of ITR per 1 mole of TER. In this arrangement, the ITR molecules are in an antiparallel orientation with a TER molecule, which is sandwiched in the formed space by the drug molecules (Figure 7). SC-XRD also identified H-bonds formed by the hydroxyl of the carboxyl acid and the nitrogen of the azole ring of ITR (Figure 7 and Table S2), in agreement with the results of vibrational spectroscopy (Figure 6). This interaction was also observed in other cocrystals of ITR [14,35], however, in the ITR–TER cocrystal, another H-bond, of the C–H ... O type, was also identified (Figure 7 and Table S2).

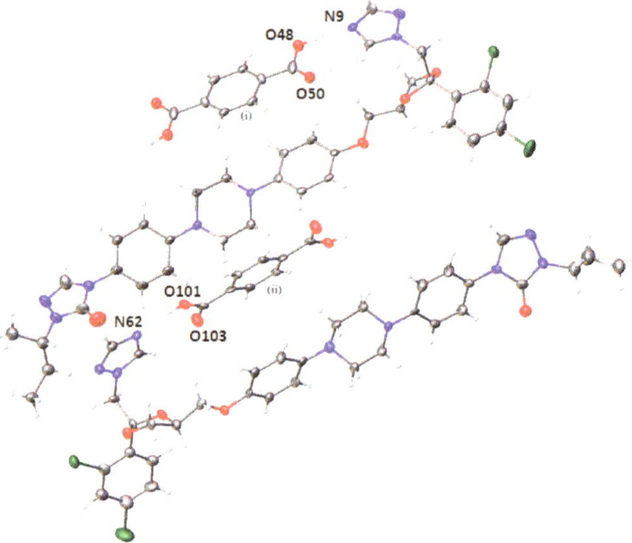

Figure 7. Crystal structure of the ITR–TER cocrystal (symmetry transformations used to generate equivalent atoms: (i) −x, 2-y, 1-z; (ii) 1-x, 1-y, 1-z) showing the major disorder in the ITR moiety only.

As the only single crystal X-ray data for ITR cocrystals available in the Cambridge Structural Database (CSD) are for ITR–succinic acid (ITR–SUC), record REWTUK [33], a more detailed comparison of the two cocrystals was made. Although both cocrystals have their crystal lattice formed by four molecules, the ITR–SUC cocrystal was described as a monoclinic system with a P2$_1$/c space group [32].

The 2D fingerprint plots, based on the 3D Hirshfeld surfaces of the ITR molecules in ITR–TER and ITR–SUC cocrystals, were generated to investigate the differences in interactions (Figure 8). It can be seen that the H ... H interactions are dominant and represent the most significant contribution (44.6% and 49.4%, in ITR–TER and ITR–SUC cocrystal, respectively) to the total Hirshfeld surfaces. In the ITR–TER cocrystal, the contribution of the C ... H contacts is greater (18.2%) in comparison to the ITR–SUC system (14.2%), most likely due to the additional weak C–H ... O H-bond. The dominant N–H ... O interaction is visible in both systems, showing as a well-defined "horn" when looking at the N ... H interactions only (Figure 8). What can be concluded is that the degree of interactions including C ... H, N ... H and O..H contacts in the ITR–TER systems is higher than that in the TER–SUC cocrystal.

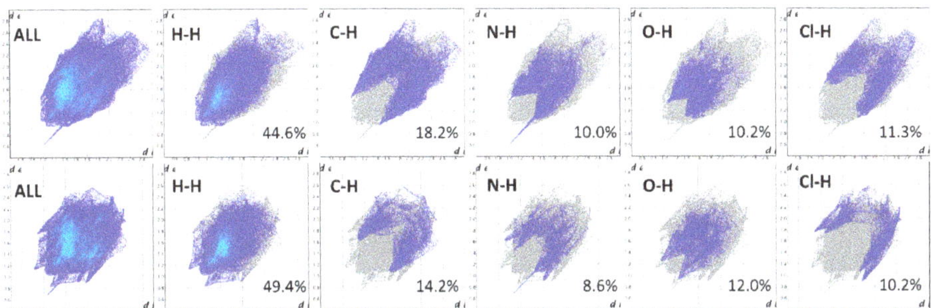

Figure 8. Two-dimensional Hirshfeld fingerprints for ITR–TER (top) and ITR–SUC (bottom) cocrystals.

3.5. Morphological Analysis

The morphological analysis revealed that the ITR powder "as received" was composed of large rod-like structures, with particles up to 30 µm (Figure 9a). An SE micrograph of particles of ITR–TER cocrystal produced by ball milling showed that this sample was composed of clusters of round particles smaller than 1 µm (Figure 9b), while the cocrystals prepared by slow solvent evaporation from methanol showed larger rod-like structures of approximately 450 µm (Figure 9c). The Bravais, Friedel, Donnay and Harker (BFDH) predicted morphology of the particle, based on the single crystal structure of ITR–TER cocrystal, showed some similarity between the experimental (Figure 9c) and simulated crystal shapes, however the (001) and (00–1) faces are clearly dominant in the former due to the solvent's impact on the crystal growth (Figure 9d).

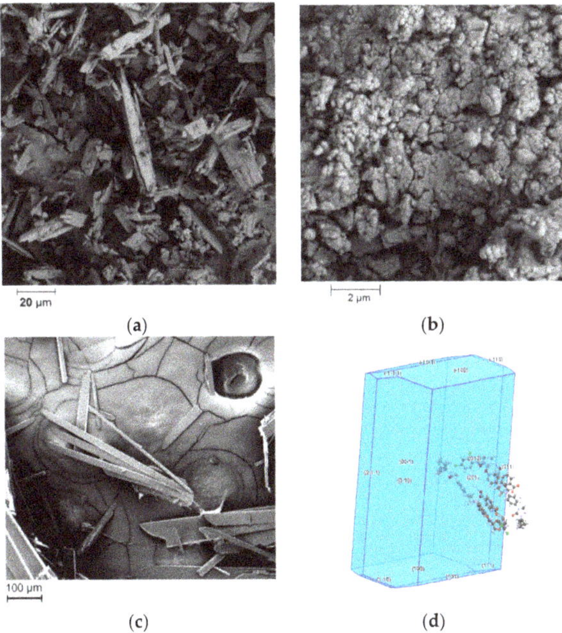

Figure 9. Scanning electron micrographs of: (**a**) crystalline ITR and the ITR–TER cocrystal produced by (**b**) ball milling, (**c**) slow solvent evaporation from methanol, as well as the (**d**) Bravais, Friedel, Donnay and Harker (BFDH) predicted morphology based on the single crystal structure of the ITR–TER cocrystal.

3.6. Pharmaceutical Evaluation of ITR Cocrystals

Following the identification, solid-state and structural characterization of the new ITR–TER cocrystal, this phase, in terms of its functional, pharmaceutical properties, was compared this those of already described cocrystals: ITR–succinic acid (ITR–SUC, [14]) and ITR–oxalic acid (ITR–OXA, [15]) as well as a freeze dried, disordered form of ITR (ITR FD). In this work, ITR–SUC and ITR–OXA cocrystals were synthesized by ball milling, so as to exclude the impact of the preparative method, and were characterized by PXRD and DSC (Figures S1 and S2) as well as SEM (Figure S3). The PXRD and DSC data for ITR FD are shown in Figure S4, while the scanning electron micrograph is shown in Figure S3.

3.6.1. Contact Angle Analysis

The contact angle of the water droplets on the surface of disks prepared by compacting ITR, FD ITR, as well as the ITR–OXA, ITR–SUC and ITR–TER cocrystals, was measured to estimate the wettability of these samples. The photos taken during this experiment showed that the shape of the water droplets remained spherical on the surface of all the disks (Figure 10), highlighting the hydrophobicity of these materials.

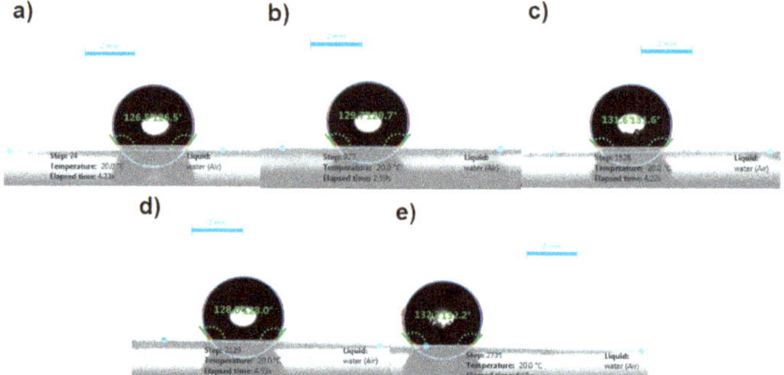

Figure 10. Representative images of water droplets on the surface of disks and contact angles of: (a) crystalline ITR, (b) freeze dried ITR (FD ITR), (c) the ITR–OXA cocrystal, (d), the ITR–SUC cocrystal and (e) the ITR–TER cocrystal.

The contact angle values of the samples were similar (Table 2), with ITR–SUC and ITR–TER showing the lowest and the highest values, respectively. The values did not change significantly after 10 min of evaluation, in comparison to the initial values, indicating that the prolonged contact of the disks with water droplets had no effect on their wettability. These results indicated that the samples investigated, regardless of the solid-state form, had poor wettability, as the values of the contact angle were greater than 90° [36].

Table 2. Contact angle of a water droplet at the start and end of the measurement as well as the statistical analysis indicating no difference between the initial and final values.

Sample	Initial (°)	Final * (°)	*p*-Value
ITR	126.8 ± 0.4	126.5 ±0.3	0.27
FD ITR	128.9 ± 0.8	127.0 ± 1.7	0.16
ITR–OXA	128.7 ± 2.9	126.0 ± 3.5	0.35
ITR–SUC	126.5 ± 1.5	124.8 ± 2.0	0.30
ITR–TER	130.5 ± 1.7	129.7 ± 1.9	0.61

(*) After 10 min of droplet deposition.

3.6.2. Intrinsic Dissolution Rate (IDR)

IDR analysis at pH 1.2 (Figure 11) revealed a different behavior of the samples, with crystalline ITR showing the slowest IDR in comparison to FD ITR, which achieved the highest dissolution rate of all the samples analyzed. The faster IDR of the FD ITR in relation to ITR was related to the disordered state of the API in the former. As identified by the solid-state characterization (Figure S4), the API in FD ITR was in the liquid crystal mesophase, a higher energy form than the crystalline state of ITR (the starting material). Therefore, the high free energy of FD ITR promoted the quicker dissolution of the API [10,37]. A one-way ANOVA revealed that the investigated samples had statistically different IDR values, except for crystalline ITR and the ITR–TER cocrystal. In this case, cocrystallisation was unable to improve the dissolution rate of the API, likely due to the poor aqueous solubility of the coformer, TER, which has an aqueous solubility of 17 µg/mL at 25 °C [38]. This value is low, but still much higher than ITR's solubility in water of around 1 ng/mL [39]. The influence of the coformer on the IDR values was evident for the other cocrystals, as the rank order of the aqueous solubility of pure coformers was OXA > SUC > TER (OXA: 130–140 mg/mL [40] and SUC: 83 mg/mL [41]), following the rank order of the IDR values of the equivalent cocrystals: ITR–OXA > ITR–SUC > ITR–TER (Figure 11). Raman spectroscopy of the disks post IDR studies showed the conservation of their solid-state characteristics.

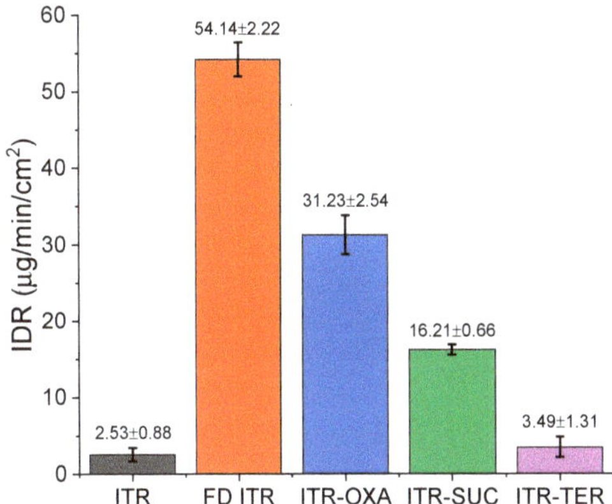

Figure 11. Intrinsic dissolution rates (IDRs) of: ITR starting material powder (ITR), freeze dried ITR (FD ITR), the ITR–OXA cocrystal, ITR–SUC cocrystal and ITR–TER cocrystal.

3.6.3. Powder/Formulation Dissolution

A powder dissolution analysis was carried out in two stages: one with the samples physically mixed with lactose monohydrate in a 1:6 *w/w* API:excipient ratio to improve the wettability and dispersibility of the drug powders, and another with the samples physically mixed with the same excipients as those in the Sporanox® formulation. In both dissolution experiments, Sporanox® capsules, a marketed formulation, was used as the reference. For the purpose of dissolution studies, the pellets were removed from the capsule shells and used without further alteration. While dissolution studies performed on ITR cocrystal particles dominate the literature [14,16], scarce information is available on the behavior during formulation, therefore the inclusion of a range of pharmaceutical excipients in the dissolution studies allows the cocrystal performance to be evaluated better.

In the dissolution experiment where the samples were mixed with lactose, the results showed that Sporanox® had a superior dissolution profile in comparison to the other samples (Figure 12a). ITR from Sporanox® had an almost constant dissolution rate until it reached its maximum percent of drug dissolved, %max, of 66.9 ± 2.1% after 40 min. In comparison to ITR, which had a %max of

3.7% ± 0.1%, the concentration of solubilized ITR released from Sporanox® was greater by 18-fold, highlighting the enhanced dissolution performance of this commercial form.

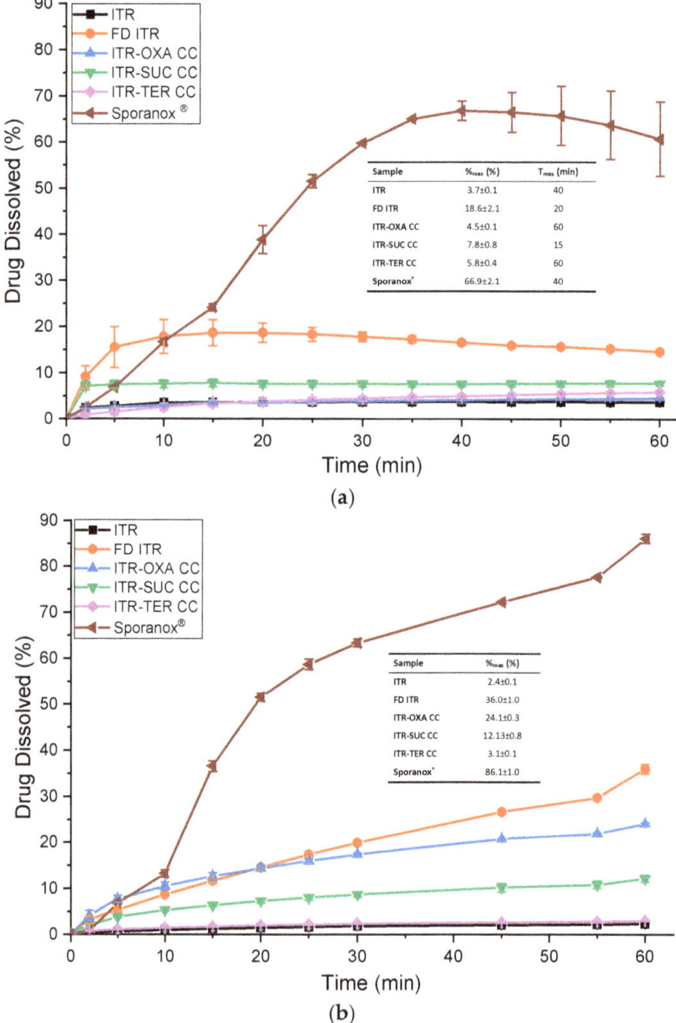

Figure 12. (**a**) Dissolution profile of Sporanox® and the physical mixtures of ITR and cocrystals with lactose monohydrate in a 1:6 ratio (*w/w*) of the active pharmaceutical ingredient (API) and excipient; (**b**) dissolution profile of Sporanox® and the ITR, FD IDR, ITR–OXA, ITR–SUC and ITR–TER mixed with excipients present in the Sporanox® capsule. T_{max}—the time to reach the maximum percent ($\%_{max}$) of the dissolved drug, CC—cocrystal.

FD ITR and ITR–SUC had faster initial dissolution rates in comparison to other ITR forms (Figure 12a). The time required to reach the maximum percent of dissolved drug, T_{max}, was 20 min for FD ITR and 15 min for ITR–SUC, while for the other samples this time was between 40 and 60 min. This effect could be due to the differences in solubility between the various forms. Nevertheless, the dissolution profile of FD ITR showed a decrease in the amount of the dissolved material from 18.6 ± 2.8% to 14.6 ± 0.4% after 60 min, which is statistically different (*t*-test, *p*-value = 0.03). This slight "parachute effect" might be caused by the crystallization of the dissolved API in order to reduce the chemical potential of the supersaturation generated when the disordered API was dissolved [42]. This was confirmed using PXRD, and it was apparent that the material remaining after the dissolution of FD ITR crystallized to form I of ITR [27].

Among the cocrystals, ITR–SUC had the highest $\%_{max}$, of 7.8% ± 0.8%, while the dissolution profile of ITR–OXA mostly overlapped with that of ITR. The cocrystal had a slightly higher $\%_{max}$, 4.5% ± 0.1%, while ITR had a $\%_{max}$ of 3.7% ± 0.1% (Figure 12a). This result contrasts with the IDR study (Figure 11), which showed a faster dissolution rate of ITR–OXA in relation to ITR. In this study, the very small difference in the percentage of the solubilized drug between ITR and ITR–OXA could be caused by an incongruent dissolution from the cocrystal and the removal of the coformer from its surface. This is due to the very different solubilities of OXA (130–140 mg/mL) in comparison to that of ITR [39,40]. A conversion to crystalline ITR form I was observed by a PXRD analysis of the remaining ITR–OXA. Therefore, the cocrystallisation of an API with a coformer with a very different aqueous solubility might not be able to enhance the dissolution rate of the API [43].

Dissolution studies of the ITR–TER cocrystal showed that the amount of the drug dissolved was increasing until the final measurement at 60 min (Figure 12a), reaching a $\%_{max}$ of 5.8% ± 0.4%, which is statistically greater by a *t*-test ($p = 0.0009$) comparison than that of crystalline ITR. The PXRD trace of the undissolved ITR–TER after the dissolution test showed no alteration in the solid-state properties after the test, suggesting that this cocrystal has an enhanced stability in aqueous media in relation to the other samples.

The dissolution analysis using the samples mixed with the same excipients as those present in the Sporanox® formulation (Figure 12b) yielded different results in comparison to the test using the physical mixtures with lactose (Figure 12a). In this experiment, Sporanox® also had a superior dissolution profile when compared to the other samples, reaching 86.1% ± 1.0% of dissolved drug. Only 2.4% ± 0.1% dissolved from crystalline ITR. FD ITR had the second highest dissolution profile. This sample continuously released ITR, reaching a $\%_{max}$ of 36.0% ± 1.0%. No "parachute" effect was observed, in contrast to the previous analysis (Figure 12a). This could be caused by the stabilizer effect of the polymers PEG 6 kDa and HPMC, preventing the crystallization of the solubilized API [44]. The PXRD trace of the undissolved FD ITR recovered after the dissolution test showed no alteration of its solid-state characteristics, suggesting that the FD ITR had an enhanced physical stability in this experiment.

In relation to the cocrystals, ITR–OXA reached the greatest $\%_{max}$ of 24.1 ± 0.3% (Figure 12b). The ITR–SUC cocrystal had a $\%_{max}$ of 12.1 ± 0.8%, while ITR–TER had the lowest but was nevertheless greater than crystalline ITR, with a drug release of 3.1 ± 0.1%. In comparison to the previous analysis, ITR–OXA and ITR–SUC showed improved dissolution profiles, probably due to the presence of polymers in the dissolution media, as observed for the FD ITR sample. The PXRD analysis of ITR–OXA and ITR–SUC after the dissolution test showed no phase change when compared to the samples before dissolution.

4. Conclusions

A new itraconazole cocrystal was produced using terephthalic acid, which is the first aromatic coformer described as being able to cocrystallise with ITR. The elucidation of the crystalline structure of the ITR–TER cocrystal revealed that this new form is similar to other known ITR cocrystals with a 2:1 stoichiometry, where the ITR molecules are in an antiparallel arrangement and the coformer is "trapped" in the space formed between the two ITR molecules. However, this supramolecular arrangement had a different heterosynthon in comparison to the existing cocrystals based on aliphatic dicarboxylic acids with an additional H-bond stabilizing the structure.

A comparison of the dissolution behavior of the various forms of ITR, including crystalline ITR, freeze dried ITR and ITR cocrystals with oxalic acid, succinic acid and terephthalic acid, showed that they had different dissolution characteristics. FD ITR showed a remarkable improvement in terms of its dissolution, considering IDR and the powder dissolution in relation to crystalline ITR. This was due to the high energy state of the API achieved by freeze drying. For the cocrystals, their IDR values and powder dissolution profiles correlated with the aqueous solubility of the coformers. The rank order of the dissolution rates of the API from the cocrystals was ITR–OXA > ITR–SUC > ITR–TER. Furthermore, physically mixing FD ITR, ITR–OXA and ITR–SUC with the same excipients as those present in the Sporanox® formulation enhanced the stability of these forms in the dissolution media, preventing their transformation into the crystalline parent API.

Supplementary Materials: The following are available online at http://www.mdpi.com/1999-4923/12/8/741/s1, Figure S1: (**a**) PXRD patterns of itraconazole (ITR) and oxalic acid (OXA) starting material powders and the ITR–OXA cocrystal produced by ball milling (BM). * Indicates distinct peaks of the cocrystal. (**b**) DSC thermograms of ITR and OXA starting material powders and the ITR–OXA cocrystal, Figure S2: (**a**) PXRD patterns of the itraconazole (ITR) and succinic acid (SUC) starting material powders and the ITR–SUC cocrystal produced by ball milling (BM). * Indicates distinct peaks of the cocrystal. (**b**) DSC thermograms of the ITR and SUC starting material powders and the ITR–SUC cocrystal, Figure S3: Scanning electron micrographs of (**a**) the ITR–SUC cocrystal and (**b**) ITR–OXA cocrystal produced by ball milling; (**c**) freeze dried ITR, Figure S4: (**a**) PXRD pattern of freeze dried itraconazole (ITR FD) and (**b**) DSC thermogram of freeze dried itraconazole (ITR FD), Table S1: Crystal data and structure refinement for the ITR–TER cocrystal, Table S2: Hydrogen bonds for ITR–TER cocrystal [Å and °]. Crystallographic data, deposition number 2015894, were deposited with the Cambridge Crystallographic Data Centre. These data can be obtained free of charge from the Cambridge Crystallographic Data Centre via www.ccdc.cam.ac.uk/structures.

Author Contributions: Conceptualization, R.M.C. and L.T.; methodology, R.M.C., T.B., J.B., B.T., O.D., L.T.; formal analysis, R.M.C., T.B., B.T., L.T.; investigation, R.M.C., T.B., J.B., B.T., M.J.S.-M., O.D., L.T.; resources, J.B., O.D., L.T.; data curation, R.M.C., T.B., E.T., B.T., O.D., L.T.; writing—original draft preparation, R.M.C., L.T.; writing—review and editing, T.B., J.B., B.T., M.J.S.-M., O.D., L.T.; visualization, R.M.C., B.T., L.T.; supervision, J.B., E.T., M.J.S.-M., O.D., L.T.; project administration, L.T.; funding acquisition, R.M.C., L.T. All authors have read and agreed to the published version of the manuscript.

Funding: This work was supported by the Coordenação de Aperfeiçoamento de Pessoal de Nível Superior (CAPES) under the agreement number 99999.001482/2015-07, the Synthesis and Solid State Pharmaceutical Centre (SSPC), financed by a research grant from Science Foundation Ireland (SFI) and co-funded under the European Regional Development Fund (grant number 12/RC/2275) and the European Union's Horizon 2020 Research and Innovation Programme under the Marie Skłodowska-Curie grant agreement No. 778051.

Conflicts of Interest: The authors declare no conflict of interest. T.B., J.B., E.T. and O.D. are employees of the Zentiva, k.s. The company had no role in the design of the study; in the collection, analyses, or interpretation of data; in the writing of the manuscript, or in the decision to publish the results.

References

1. Schultheiss, N.; Newman, A. Pharmaceutical cocrystals and their physicochemical properties. *Cryst. Growth Des.* **2009**, *9*, 2950–2967. [CrossRef]

2. Thakuria, R.; Delori, A.; Jones, W.; Lipert, M.P.; Roy, L.; Rodríguez-Hornedo, N. Pharmaceutical cocrystals and poorly soluble drugs. *Int. J. Pharm.* **2013**, *453*, 101–125. [CrossRef]

3. Berry, D.J.; Steed, J.W. Pharmaceutical cocrystals, salts and multicomponent systems; intermolecular interactions and property based design. *Adv. Drug Deliv. Rev.* **2017**, *117*, 3–24. [CrossRef]

4. Desiraju, G.R. Supramolecular Synthons in Crystal Engineering—A New Organic Synthesis. *Angew. Chem. Int. Ed. Engl.* **1995**, *34*, 2311–2327. [CrossRef]

5. Cerreia Vioglio, P.; Chierotti, M.R.; Gobetto, R. Pharmaceutical aspects of salt and cocrystal forms of APIs and characterization challenges. *Adv. Drug Deliv. Rev.* **2017**, *117*, 86–110. [CrossRef]

6. Campoy, S.; Adrio, J.L. Antifungals. *Biochem. Pharmacol.* **2017**, *133*, 86–96. [CrossRef] [PubMed]

7. Denning, D.W.; Hope, W.W. Therapy for fungal diseases: Opportunities and priorities. *Trends Microbiol.* **2010**, *18*, 195–204. [CrossRef] [PubMed]

8. Papadopoulou, V.; Valsami, G.; Dokoumetzidis, A.; Macheras, P. Biopharmaceutics classification systems for new molecular entities (BCS-NMEs) and marketed drugs (BCS-MD): Theoretical basis and practical examples. *Int. J. Pharm.* **2008**, *361*, 70–77. [CrossRef] [PubMed]

9.   Hardin, T.C.; Graybill, J.R.; Fetchick, R.; Woestenborghs, R.; Rinaldi, M.G.; Kuhn, J.G. Pharmacokinetics of itraconazole following oral administration to normal volunteers. *Antimicrob. Agents Chemother.* **1988**, *32*, 1310–1313. [CrossRef] [PubMed]

10.   Mugheirbi, N.A.; Paluch, K.J.; Tajber, L. Heat induced evaporative antisolvent nanoprecipitation (HIEAN) of itraconazole. *Int. J. Pharm.* **2014**, *471*, 400–411. [CrossRef] [PubMed]

11.   Mugheirbi, N.A.; Tajber, L. Mesophase and size manipulation of itraconazole liquid crystalline nanoparticles produced via quasi nanoemulsion precipitation. *Eur. J. Pharm. Biopharm.* **2015**, *96*, 226–236. [CrossRef] [PubMed]

12.   Kozyra, A.; Mugheirbi, N.A.; Paluch, K.J.; Garbacz, G.; Tajber, L. Phase Diagrams of Polymer-Dispersed Liquid Crystal Systems of Itraconazole/Component Immiscibility Induced by Molecular Anisotropy. *Mol. Pharm.* **2018**, *15*, 5192–5206. [CrossRef] [PubMed]

13.   Kalepu, S.; Nekkanti, V. Insoluble drug delivery strategies: Review of recent advances and business prospects. *Acta Pharm. Sin. B* **2015**, *5*, 442–453. [CrossRef] [PubMed]

14.   Remenar, J.F.; Morissette, S.L.; Peterson, M.L.; Moulton, B.; MacPhee, J.M.; Guzmán, H.R.; Almarsson, Ö. Crystal engineering of novel cocrystals of a triazole drug with 1,4-dicarboxylic acids. *J. Am. Chem. Soc.* **2003**, *125*, 8456–8457. [CrossRef] [PubMed]

15.   Shevchenko, A.; Miroshnyk, I.; Pietilä, L.O.; Haarala, J.; Salmia, J.; Sinervo, K.; Mirza, S.; Van Veen, B.; Kolehmainen, E.; Nonappa; et al. Diversity in itraconazole cocrystals with aliphatic dicarboxylic acids of varying chain length. *Cryst. Growth Des.* **2013**, *13*, 4877–4884. [CrossRef]

16.   Weng, J.; Wong, S.N.; Xu, X.; Xuan, B.; Wang, C.; Chen, R.; Sun, C.C.; Lakerveld, R.; Kwok, P.C.L.; Chow, S.F. Cocrystal Engineering of Itraconazole with Suberic Acid via Rotary Evaporation and Spray Drying. *Cryst. Growth Des.* **2019**, *19*, 2736–2745. [CrossRef]

17.   Ma, Y.H.; Ge, S.W.; Wang, W.; Sun, B.W. Studies on the synthesis, structural characterization, Hirshfeld analysis and stability of apovincamine (API) and its co-crystal (terephthalic acid: Apovincamine = 1:2). *J. Mol. Struct.* **2015**, *1097*, 87–97. [CrossRef]

18.   Mikhailovskaya, A.V.; Myz, S.A.; Bulina, N.V.; Gerasimov, K.B.; Kuznetsova, S.A.; Shakhtshneider, T.P. Screening and characterization of cocrystal formation between betulin and terephthalic acid. *Mater. Today Proc.* **2020**, *25*, 381–383. [CrossRef]

19.   Lemmerer, A.; Bernstein, J.; Kahlenberg, V. Hydrogen bonding patterns of the co-crystal containing the pharmaceutically active ingredient isoniazid and terephthalic acid. *J. Chem. Crystallogr.* **2011**, *41*, 991–997. [CrossRef]

20.   André, V.; Fernandes, A.; Santos, P.P.; Duarte, M.T. On the track of new multicomponent gabapentin crystal forms: Synthon competition and pH stability. *Cryst. Growth Des.* **2011**, *11*, 2325–2334. [CrossRef]

21.   Rodríguez-Spong, B.; Price, C.P.; Jayasankar, A.; Matzger, A.J.; Rodríguez-Hornedo, N. General principles of pharmaceutical solid polymorphism: A supramolecular perspective. *Adv. Drug Deliv. Rev.* **2004**, *56*, 241–274. [CrossRef] [PubMed]

22.   Paluch, K.J.; Tajber, L.; McCabe, T.; O'Brien, J.E.; Corrigan, O.I.; Healy, A.M. Preparation and solid state characterisation of chlorothiazide sodium intermolecular self-assembly suprastructure. *Eur. J. Pharm. Sci.* **2010**, *41*, 603–611. [CrossRef] [PubMed]

23.   Machado Cruz, R.; Santos-Martinez, M.J.; Tajber, L. Impact of polyethylene glycol polymers on the physicochemical properties and mucoadhesivity of itraconazole nanoparticles. *Eur. J. Pharm. Biopharm.* **2019**, *144*, 57–67. [CrossRef] [PubMed]

24.   Turner, M.J.; McKinnon, J.; Wolff, S.K.; Grimwood, D.J.; Spackman, P.R.; Jayatilaka, D.; Spackman, M.A. CrystalExplorer 17. 2017. Available online: http://hirshfeldsurface.net (accessed on 1 July 2020).

25.   Frišić, T.; Childs, S.L.; Rizvi, S.A.A.; Jones, W. The role of solvent in mechanochemical and sonochemical cocrystal formation: A solubility-based approach for predicting cocrystallisation outcome. *CrystEngComm* **2009**, *11*, 418–426. [CrossRef]

26.   Braga, D.; Maini, L.; Grepioni, F. Mechanochemical preparation of co-crystals. *Chem. Soc. Rev.* **2013**, *42*, 7638–7648. [CrossRef]

27.   Zhang, S.; Lee, T.W.Y.; Chow, A.H.L. Crystallization of Itraconazole Polymorphs from Melt. *Cryst. Growth Des.* **2016**, *16*, 3791–3801. [CrossRef]

28. Karagianni, A.; Malamatari, M.; Kachrimanis, K. Pharmaceutical cocrystals: New solid phase modification approaches for the formulation of APIs. *Pharmaceutics* **2018**, *10*, 18. [CrossRef]

29. Elmas Kimyonok, A.B.; Ulutürk, M. Determination of the Thermal Decomposition Products of Terephthalic Acid by Using Curie-Point Pyrolyzer. *J. Energ. Mater.* **2016**, *34*, 113–122. [CrossRef]

30. Yamashita, H.; Hirakura, Y.; Yuda, M.; Terada, K. Coformer screening using thermal analysis based on binary phase diagrams. *Pharm. Res.* **2014**, *31*, 1946–1957. [CrossRef]

31. Williams, D.H.; Fleming, I. *Spectroscopic Methods in Organic Chemistry*, 5th ed.; McGraw-Hill Education: New York, NY, USA, 1995; ISBN 9780077091477.

32. Karthikeyan, N.; Joseph Prince, J.; Ramalingam, S.; Periandy, S. Electronic [UV–Visible] and vibrational [FT-IR, FT-Raman] investigation and NMR–mass spectroscopic analysis of terephthalic acid using quantum Gaussian calculations. *Spectrochim. Acta Part A Mol. Biomol. Spectrosc.* **2015**, *139*, 229–242. [CrossRef]

33. Lahtinen, M.; Kolehmainen, E.; Haarala, J.; Shevchenko, A. Evidence of weak halogen bonding: New insights on itraconazole and its succinic acid cocrystal. *Cryst. Growth Des.* **2013**, *13*, 346–351. [CrossRef]

34. Delori, A.; Frišić, T.; Jones, W. The role of mechanochemistry and supramolecular design in the development of pharmaceutical materials. *CrystEngComm* **2012**, *14*, 2350–2362. [CrossRef]

35. Shevchenko, A.; Bimbo, L.M.; Miroshnyk, I.; Haarala, J.; Jelínková, K.; Syrjänen, K.; Van Veen, B.; Kiesvaara, J.; Santos, H.A.; Yliruusi, J. A new cocrystal and salts of itraconazole: Comparison of solid-state properties, stability and dissolution behavior. *Int. J. Pharm.* **2012**, *436*, 403–409. [CrossRef]

36. Florence, A.T.; Attwood, D. *Physicochemical Principles of Pharmacy*, 4th ed.; Pharmaceutical Press: London, UK, 2006.

37. Mugheirbi, N.A.; O'Connell, P.; Serrano, D.R.; Healy, A.M.; Taylor, L.S.; Tajber, L. A Comparative Study on the Performance of Inert and Functionalized Spheres Coated with Solid Dispersions Made of Two Structurally Related Antifungal Drugs. *Mol. Pharm.* **2017**, *14*, 3718–3728. [CrossRef] [PubMed]

38. Park, C.-M.; Sheehan, R.J. Phthalic Acids and Other Benzenepolycarboxylic Acids. In *Kirk-Othmer Encyclopedia of Chemical Technology*; John Wiley & Sons Inc.: Hoboken, NJ, USA, 2000.

39. Stevens, D. A Itraconazole in Cyclodextrin Solution. *Pharmacotherapy* **1999**, *19*, 603–611. [CrossRef]

40. Tyner, T.; Francis, J. Oxalic Acid Dihydrate. In *ACS Reagent Chemicals*; American Chemical Society: Washington, DC, USA, 2018.

41. Litsanov, B.; Brocker, M.; Oldiges, M.; Bott, M. Succinic Acid. In *Bioprocessing of Renewable Resources to Commodity Bioproducts*; John Wiley & Sons Inc.: Hoboken, NJ, USA, 2014; pp. 435–472. ISBN 9781118845394.

42. Babu, N.J.; Nangia, A. Solubility Advantage of Amorphous Drugs and Pharmaceutical Cocrystals. *Cryst. Growth Des.* **2011**, *11*, 2662–2679. [CrossRef]

43. Grossjohann, C.; Eccles, K.S.; Maguire, A.R.; Lawrence, S.E.; Tajber, L.; Corrigan, O.I.; Healy, A.M. Characterisation, solubility and intrinsic dissolution behaviour of benzamide: Dibenzyl sulfoxide cocrystal. *Int. J. Pharm.* **2012**, *422*, 24–32. [CrossRef]

44. Van Eerdenbrugh, B.; Van den Mooter, G.; Augustijns, P. Top-down production of drug nanocrystals: Nanosuspension stabilization, miniaturization and transformation into solid products. *Int. J. Pharm.* **2008**, *364*, 64–75. [CrossRef]

 *pharmaceutics*

Article

Salt Cocrystal of Diclofenac Sodium-L-Proline: Structural, Pseudopolymorphism, and Pharmaceutics Performance Study

Ilma Nugrahani [1,*], Rizka A. Kumalasari [1], Winni N. Auli [1], Ayano Horikawa [2] and Hidehiro Uekusa [2]

[1] School of Pharmacy, Bandung Institute of Technology, Bandung, West Java 40132, Indonesia; kumalsarizka@gmail.com (R.A.K.); winnie.nur.auly@gmail.com (W.N.A.)
[2] Department of Chemistry, School of Science, Tokyo Institute of Technology, Tokyo 152-8551, Japan; a.horikawa.chem@gmail.com (A.H.); uekusa@chem.titech.ac.jp (H.U.)
* Correspondence: ilma_nugrahani@fa.itb.ac.id; Tel./Fax: +62-22 2504852

Received: 17 June 2020; Accepted: 12 July 2020; Published: 21 July 2020

Abstract: Previously, we have reported on a zwitterionic cocrystal of diclofenac acid and L-proline. However, the solubility of this multicomponent crystal was still lower than that of diclofenac sodium salt. Therefore, this study aimed to observe whether a multicomponent crystal could be produced from diclofenac sodium hydrate with the same coformer, L-proline, which was expected to improve the pharmaceutics performance. Methods involved screening, solid phase characterization, structure determination, stability, and in vitro pharmaceutical performance tests. First, a phase diagram screen was carried out to identify the molar ratio of the multicomponent crystal formation. Next, the single crystals were prepared by slow evaporation under two conditions, which yielded two forms: one was a rod-shape and the second was a flat-square form. The characterization by infrared spectroscopy, thermal analysis, and diffractometry confirmed the formation of the new phases. Finally, structural determination using single crystal X-ray diffraction analysis solved the new salt cocrystals as a stable diclofenac–sodium–proline–water (1:1:1:4) named NDPT (natrium diclofenac proline tetrahydrate), and unstable diclofenac–sodium–proline–water (1:1:1:1), named NDPM (natrium diclofenac proline monohydrate). The solubility and dissolution rate of these multicomponent crystals were superior to those of diclofenac sodium alone. The experimental results that this salt cocrystal is suitable for further development.

Keywords: diclofenac sodium; L-proline; salt cocrystal; multicomponent crystal; monohydrate; tetrahydrate; solubility; dissolution; stability

---

1. Introduction

Diclofenac is a well-established non-steroidal anti-inflammatory drug (NSAID) with anti-inflammatory, analgesic, and antipyretic activities [1], and is the most widely consumed NSAID in some countries. In Indonesia, diclofenac is one of the three most widely used NSAID drugs and represents 14.4% of all NSAIDs consumed in the country [2]. Based on the Biopharmaceutics Classification System (BCS), diclofenac is a class II compound with high permeability but low solubility, which results in limited bioavailability. Therefore, it is necessary to increase the solubility of the compound.

Techniques that can improve the solubility and bioavailability of drugs include cosolvation, surfactant usage, pH adjustment, solid dispersion, salt formation, and cocrystallization [3–7]. Pharmaceutical cocrystals are multicomponent systems in which two or more components, in a drug–drug or drug–coformer combination, are present in a stoichiometric ratio and bonded together with hydrogen interactions in a crystal lattice [7]. The rationale behind screening for suitable coformers for pharmaceutical development of a hydrophobic drug should consider the rule of five by Lapinski, i.e., hydrogen bonding, halogen bonding (and non-covalent bonding in general), length of the carbon chain, molecular recognition points, and aqueous solubility [8,9].

Cocrystals can be classified into three groups: anhydrous/non-solvated cocrystal, hydrate/solvate cocrystal, and salt cocrystal [9–11]. We are focusing on salt cocrystallization, which is a special kind of cocrystal having a salt structure (anion and cation) with a neutral coformer. It is expected to have a combined benefit of the salt and cocrystal properties, such as improved stability, solubility, dissolution, as well as enhanced the bioavailability [12–16]. As previously reported, salt cocrystal niclosamide and salt cocrystal pefloxacin can enhance solubility and dissolution [17,18]. Moreover, our previous research has shown that a salt cocrystal can be made from saccharine sodium salt with theophylline, as well as acidic saccharine with a similar excipient [19].

Diclofenac acid is a poorly soluble drug. Recently, we have attempted and reported on a cocrystal of neutral diclofenac acid with a zwitterionic coformer, L-proline (LP), which increased the drug solubility by more than 7.6-fold [20]. However, the solubility level was not improved over the salt form, diclofenac sodium, which has been developed and more commonly used in dosage forms at the present time [21–29]. Some reports have described that this sodium salt can be found in anhydrous and several hydrated forms, i.e., the tetrahydrate [24–26], pentahydrate [27], trihydrate [28], and 4.75-hydrate [29], respectively. These phenomena that cause commercial of anhydrous diclofenac sodium (ND) are often found in the mixture with the hydrate phases (NDH) during distribution and storage [28]. Fortunately, NDH could change back to ND by heating and drying [25,26,28,29].

This study aimed to discover salt cocrystals of NDH with LP (abbreviated NDP) to obtain the drug's superior properties, namely, stability, solubility, and dissolution [3–7,14–20]. Considering that ND was unstable, we initially arranged a cocrystal from NDH to develop a suitable and practical method, which can be performed readily even under the ambient conditions. As LP is a zwitterion, it was not expected to form a salt with diclofenac acid moiety. The cationic and anionic groups in this amino acid were predicted to arrange strong hydrogen bonds or charge interactions with other molecules [20]. The most important of LP physicochemical aspects are its high wettability, hydrotropic property, and its flexibility to dissolve in a wide pH range [20,30–32]. Besides, LP was selected again in this work due to its safety, economy, and availability.

The experiment began with screening the physical interaction between the compounds using a phase diagram, and then the cocrystal was isolated utilizing a specific molar ratio of components. The cocrystal was characterized by binocular microscope, infrared spectroscopy, thermal analysis, and powder X-ray diffractometry, then it was assessed by single crystal X-ray diffractometry until the final 3D structure was determined. Due to the pseudopolymorphism phenomenon of this new multicomponent cocrystal, the parameters that need to be discussed are the hydrate stability, which was tested by drying and humidifying at specific temperature and humidity levels. The next steps were solubility and dissolution tests. The results of this study are expected to demonstrate that this new phase derived from ND and LP increases the solubility and dissolution of ND and is stable.

2. Materials and Methods

2.1. Materials

Materials used in this experiment were commercial pharmaceutical-grade of ND (Pharos Indonesia, Semarang, Indonesia); ND-anhydrous pro-analysis 99.9% from Sigma Aldrich (Jakarta, Indonesia); L-proline from Tokyo Chemical Industry (Tokyo, Japan); methanol, ethanol, acetone, and potassium bromide (KBr) crystal for infrared analysis; sodium hydroxide (NaOH); potassium hydrogen phosphate (KH_2PO_4); and hydrochloride acid (HCl); these reagents, solvents, and materials were purchased from Sigma Aldrich and Merck (Jakarta, Indonesia). Distilled water was prepared by Bandung Institute of Technology (Bandung, Indonesia) and aluminum pans were obtained from Rigaku (Tokyo, Japan).

2.2. Methods

2.2.1. Preparation and Characterization of Starting Materials

The references have stated that ND can easily change to NDH, which was stable in room temperature and atmosphere, causing commercial ND is commonly found in mixture with its hydrates [25,26,28,29]. Therefore, this experiment was started by preparing a homogeneous phase of NDH to develop a method under ambient conditions (condition I) promptly, besides under the restricted low humidity (condition II). NDH was prepared by storing commercial ND (Pharos Indonesia Ltd., Semarang, Indonesia, claimed as anhydrous) in an opened container under ambient temperature (72 ± 2% RH/25 ± 2 °C) for 24 h. Anhydrous ND was purified by storing the commercial ND in a desiccator with silica gel for 24 h for cocrystal preparation under the environment with the minimal water (condition II). NDH, ND, and LP were characterized using a Fourier transform infrared (FTIR) spectrophotometry, differential scanning calorimetry (DSC), differential thermal analysis/thermogravimetry (DTA/TG), and powder X-ray diffractometry (PXRD).

2.2.2. Screening to Determine the Optimal Molar Ratio of Cocrystals Using a Binary Phase Diagram

The physical mixtures of NDH and LP were made in several molar ratios of 10:0, 9:1, 8:2, 7:3, 6:4, 5:5, 4:6, 3:7, 2:8, 1:9, and 0:10, or 0.0–1.0 molar fraction of NDH in the mixture with LP. The ingredients were weighed using a Mettler Toledo microbalance (Greifensee, Switzerland) and mixed gently (to avoid the cocrystal formation in situ) until homogeneous. For each composition, the transition phase temperature was measured using an electrothermal, and the melting point data were plotted against the fraction molar of NDH to make a phase diagram [33–36].

2.2.3. Preparation of NDP Cocrystals

The cocrystal preparation was performed using two conditions. First, ~1 g mixture of NDH and LP in 1:1 molar ratio was dissolved in 25 mL of various solvents: ethanol, acetone, and methanol (80–90%), respectively (condition I). The clear solutions were then slowly evaporated under ambient conditions (75 ± 2% RH/25 ± 2 °C). Second, a similar amount of preparation was made from anhydrous ND - LP in a 1:1 molar ratio, which was dissolved in 25 mL volume of the purer ethanol, acetone, and methanol (90–95%) (condition II). The cocrystallization was performed in a fume hood under room temperature.

2.2.4. Electrothermal Measurement

A small amount of sample was filled into a capillary tube and then was put on the sample holder of the Electrothermal AZ 9003 apparatus (Staffordshire, UK). The start temperature on the device was set at 10 °C under the first transition temperature point of each sample, with a heating rate of 10 °C min^{-1}. Phase changes, including water releasing, melting point, and decomposition, were observed via the magnification viewer on the apparatus. The temperature related to the sample's behavior under heating was recorded thoroughly.

2.2.5. Crystal Habit Observation

The habit of the recrystallized NDH, LP, and NDP from Condition I and II were put on the objective glass and were observed under a binocular microscope Olympus CX21, (Tokyo, Japan), without cover glass. The pictures were taken by I-phone 7 camera in magnification 100×.

2.2.6. FTIR Measurement

The amount of 99 mg KBr crystal and 1 mg sample were mixed and compressed to be a disc using a hydraulic presser. The pellet was mounted on the holder, and the infrared spectral measurements were carried out using a FTIR Jasco 4200 Type-A (Easton, PA, USA), over the wavenumber range of

4000 to 400 cm^{-1} at 4 cm^{-1} resolution. The starting materials (NDH and LP), physical mixture 1:1 (PM), and salt cocrystals (NDP) were analyzed. Data were processed by Microsoft Excel 2003 and Origin-Pro 8.5.1 software. The second derivative spectra were generated to enhance the specificity [37,38].

2.2.7. DSC Measurement

DSC measurement was performed using a Rigaku Thermoplus DSC 8230 (Tokyo, Japan) under a nitrogen purge of 50 mL/min. For this, 3–5 mg of sample was placed in a closed aluminum pan and measured in a temperature range of 30 to 350 °C for ND and 30 to 250 °C for LP and the cocrystals, with a heating speed of 10 °C min^{-1}. An empty closed aluminum was used as a reference.

2.2.8. DTA/TG Measurement

DTA/TG measurement was performed under a nitrogen purge of 50 mL/min using a Rigaku Thermoplus DTA/TG 8122 apparatus (Tokyo, Japan). For this, 5–10 mg of sample was placed into an open aluminum pan and measured in a temperature range of 30 to 350 °C for NDH and 30 to 250 °C for LP and the cocrystals. An empty open aluminum pan was used as a reference. All measurements were conducted with a heating speed of 10 °C min^{-1}.

2.2.9. PXRD Measurement

The sample powder was placed between Mylar$^{(R)}$ film on sample holder of Rigaku SmartLab PXRD (Tokyo, Japan). The powder diffraction pattern was collected from 2θ = 3° to 40° at ambient temperature at the step and scan speeds of 0.01° and 3° min^{-1}, respectively, using a Cu-Kα source at 45 kV and 200 mA. The PXRD pattern was plotted using Microsoft Excel 2003 and Origin-Pro 8.5.1 software.

2.2.10. Crystal Structure Analysis Using Single-Crystal X-ray Diffraction Analysis (SCXRD)

Single-crystal X-ray diffraction data were collected in ω-scan mode using a Rigaku R-AXIS RAPID diffractometer (Tokyo, Japan) with a Mo Kα radiation (λ = 0.71075 Å) rotating anode source. The integrated and scaled data were empirically corrected for absorption effects using ABSCOR. The initial structures were solved using direct methods with SHELXT and refined with SHELXL. All nonhydrogen atoms were refined anisotropically. All hydrogen atoms were found in a difference Fourier map; however, they were placed by geometrical calculations and treated using a riding model during the refinement. Two salt cocrystals of diclofenac sodium–proline–water were found and determined successfully, which then were named NDPT (from condition I) and NDPM (from condition II).

2.2.11. Stability Test

NDPT stability was tested under three conditions. The first treatment was drying in a controllable incubator Eyela (Tokyo, Japan), which was set at 30 ± 0.5% RH/40 ± 0.5 °C to observe water release. Sample was taken from the incubator after 0, 2, 4, 6, 12, and 24 h of drying. The second treatment consisted of the NDPT sample at room humidity and temperature (72 ± 2% RH/25 ± 2 °C), which was sampled periodically until 72 h. The third stability test was performed by storing the dried sample at high relative humidity (~94% RH), which was prepared by filling a chamber with a saturated solution of potassium nitrate [16]. Sampling was carried out periodically to investigate the changes of water molecules and hydrate stability for 15 days by DTA/TG, DSC, and PXRD.

2.2.12. Solubility Test

Previously, a calibration curve was composed of a series of concentration of anhydrous ND standard solutions in CO_2-free distilled water (pH 7.0), which measured by a Beckman Coulter DU720 UV–Vis spectrophotometer (Miami, FL, USA) at a wavelength of 276 nm [1,20,26]. Samples of ND, the PM, NDPM, and NDPT were placed into a 50 mL Erlenmeyer flask, and 10 mL of CO_2 free distilled water was added. The mixture was stirred on Oregon KJ-201BD shaker (Jiangsu, China) for 8 h at a speed of 100 rpm at 25 ± 2 °C, and the steady saturated equilibrium solubility was maintained. The concentrations of the sample were determined using absorbance conversion through a previously created regression equation.

2.2.13. Dissolution Test

Testing began with making dissolution medium mirroring conditions in the human stomach (pH 1.2) and intestine (pH 6.8). The pH 1.2 medium was prepared from 16.67 mL HCl—12 M, which was added to distilled water (pH 7.0) and was then adjusted to 1L. It was fixed at a pH of 1.2.

The "intestineal" dissolution medium was made from aqueous KH_2PO_4 0.2 M with NaOH, which was dissolved in 250 mL of CO_2-free distilled water (pH 7.0). This mixture was then added with distilled water until 1 L and adjusted with NaOH and H_3PO_4 to a fixed pH of 6.8. The pH values were checked using a pH meter from Mettler Toledo (Jakarta, Indonesia).

A calibration curve for concentration determination was composed previously of a series of standard solutions of ND (anhydrous-pro analysis from Sigma Aldrich) in each medium. The absorbance was measured at $\lambda = 276$ nm using the UV–visible spectrophotometer.

Before testing, the diclofenac level of all samples was calculated based on the UV absorbance comparison referred to as the ND standard reference. The equality value was used for weighing the samples, which must be equivalent to 200 mg anhydrous ND. Each powder sample for dissolution testing was sifted using a 120 mesh sieve and then was placed in the basket of Guoming RC-1 dissolution apparatus (Shanghai, China), dipped in the dissolution flask containing 900 mL of medium at a temperature of 37 ± 0.5 °C, and rotated at a speed of 50 rpm. A 2 mL volume of sample was taken after 3, 5, 10, 15, 30, and 45 min. Each sample solution was previously filtered through a 0.45 μm membrane filter before the UV absorbance measurement.

3. Results and Discussion

3.1. ND and NDH Preparation

The hydrate forms of ND have been extensively reported [24–29]. In this experiment, NDH was used as the initial material due to its stability compared to anhydrous ND, as previously stated, to absorbed water from the environment quickly. The instability of anhydrous ND raw material in pharmaceutical manufacturing caused it to mix with the hydrate forms. However, the dehydrated material can be obtained by heating or drying of NDH under specific conditions [25,26,28,29]. Therefore, we considered using NDH as the first starting material for the cocrystal screening instead of ND. First, NDH was prepared by storing the commercial ND (claimed as anhydrous) under ambient conditions ($72 \pm 2\%$ RH/25 ± 2 °C) for 24 h.

As has been cited in Introduction section, there are some reported hydrate forms. However, FTIR, PXRD, and TG data confirmed that the NDH obtained from this technique was a tetrahydrate, which consisted of 20% water [26,28]. We also checked that even NDH changed to ND by restoring it in the desiccator with silica gel for 24 h (referred to anhydrous ND analysis standard, 99.9%, from Sigma Aldrich, Jakarta, Indonesia), also met the reports [25,26,28]. The NDH has been rechecked to be stable during the experiment, following the works in [25,26,29]. In addition, we also observed that NDH could be dehydrated and back to ND entirely by storing it in a desiccator with silica gel during 24 h. Powder X-ray diffractogram and TG thermogram data of NDH and ND comparing to anhydrous ND analysis standard are attached in Supplementary S1–S3.

3.2. Determination of Molar Ratio in NDP for Cocrystallization

The fixed molar ratio of ND and LP for arranging the cocrystal was determined using thermal profile analysis. Modification of covalent bonds in a chemical structure can change its melting behavior.

Within the survey, 50 cocrystal samples were analyzed, and 51% of the cocrystals analyzed had melting points between those of the active pharmaceutical ingredient (API) and coformer, while 39% were lower than either the API or coformer. Only 6% had melting points higher than the starting materials, and 4% had melting points equal to either the API or coformer [15]. Thermal analysis through the construction of melting temperature versus the composition on a phase diagram is known to establish whether a combination forms a cocrystal or eutectic [15,33–36]. A typical binary phase diagram of a eutectic mixture will depict a V shape with one eutectic point. A cocrystal forms a "W" shape, with two eutectic points, and one higher melting point between them.

The melting temperature data (*y*-axis) of NDH–LP mixtures were plotted against the NDH molar fraction (*x*-axis) to produce a phase diagram (Figure 1). The phase diagram showing a "W" cocrystal typical pattern, with two eutectic points found at the molar fractions of ND 0.3 (at 88 °C) and 0.6 (at 98 °C), respectively. The graphic reveals that the 0.5 molar fraction of ND is the peak, has a higher melting temperature (at 116 °C) than the eutectic points. Therefore, NDH–LP (abbreviated NDP system) could be predicted to form a cocrystal with the 1:1 molar ratio as the fixed stoichiometric proportion.

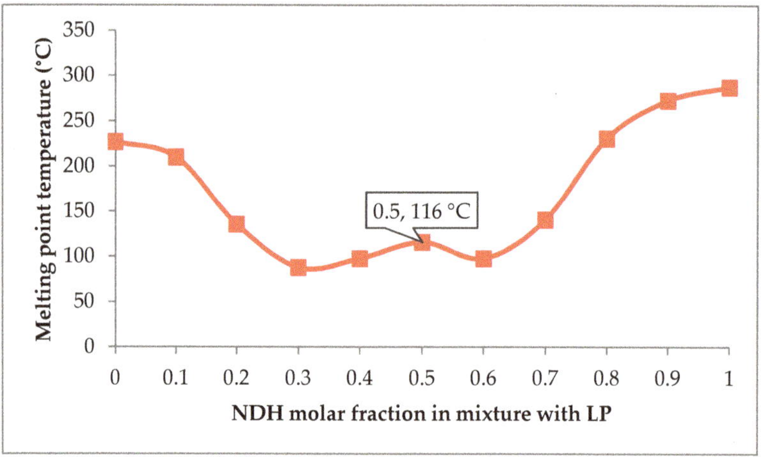

Figure 1. Phase diagram of diclofenac sodium hydrate (NDH) and L-proline (LP) composed from a series of molar ratios, represented by the molar fraction of NDH in the mixture towards the melting point.

3.3. NDP Cocrystals Preparation and Characterization

3.3.1. Crystal Habit Observation

The crystal morphology was observed as simple identification, which then should be analyzed thoroughly with the more precise solid analysis instruments. Based on the screening data, cocrystals were collected by dissolving the (1:1) molar ratio of NDH–LP mixture in the first condition explained in the Methods section. It was observed that the crystallization of NDP from all solvents occurred fast (less than 24 h) under room temperature and yielded a similar habit and different from the starting materials. As a visualization, Figure 2 shows the crystal picture of the rectangular NDH (a), needle-shaped LP (b), and a cylindrical rods-shaped NDP (c). The different crystal habit of diclofenac–sodium–proline (NDP) from the starting material may indicate the new phase formation, which then should be confirmed thoroughly, by solid characterization using FTIR, PXRD, DSC/DTA/TG, and SCXRD.

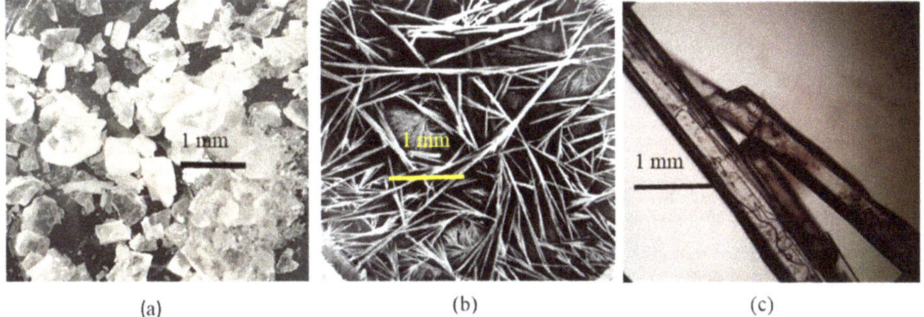

(a)   (b)   (c)

Figure 2. Crystal morphology under a binocular microscope of (**a**) diclofenac sodium hydrate (NDH), (**b**) L-proline (LP), and (**c**) diclofenac–sodium–proline cocrystal (NDP). All recrystallizations were used ethanol solvent 95% under ambient conditions (72 ± 2% RH/25 ± 2 °C).

3.3.2. FTIR Measurement Data

In the crystal arrangement, FTIR has commonly used to detect the hydrogen bonding formation [37–45]. Figure 3a reveals infrared spectra of NDH, LP, physical mixture 1:1 (PM), and NDP crystals from ethanol, methanol, and acetone, respectively. FTIR spectra of NDH indicated the water –OH stretching band at 3200–3500 cm^{-1} and COO$^-$ stretching band at 1631 and 1573 cm^{-1} [21,26]. LP depicts the C=O stretching at 1619 cm^{-1} and COO$^-$ stretching of amino band at 1400 cm^{-1} [20,37]. Next, similar spectra were shown by NDP cocrystals obtained from all solvents (methanol, ethanol, and acetone). It can be seen that the cocrystal spectra were different from PM in several bands between 4000 and 400 cm^{-1} regions of measurement.

Due to some overlapped bands, the second derivative spectra (Figure 3b) was generated to increase specificity by separating the overlapped bands on the similar region [37,38]. There are three focused areas shown in Figure 3, with the distinctive bands assigned by the wavenumber point and arrow.

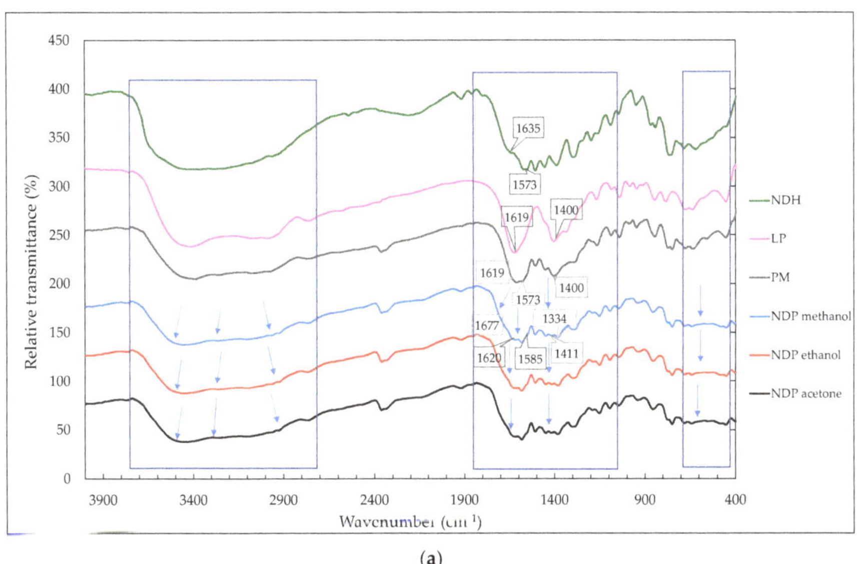

(a)

Figure 3. *Cont.*

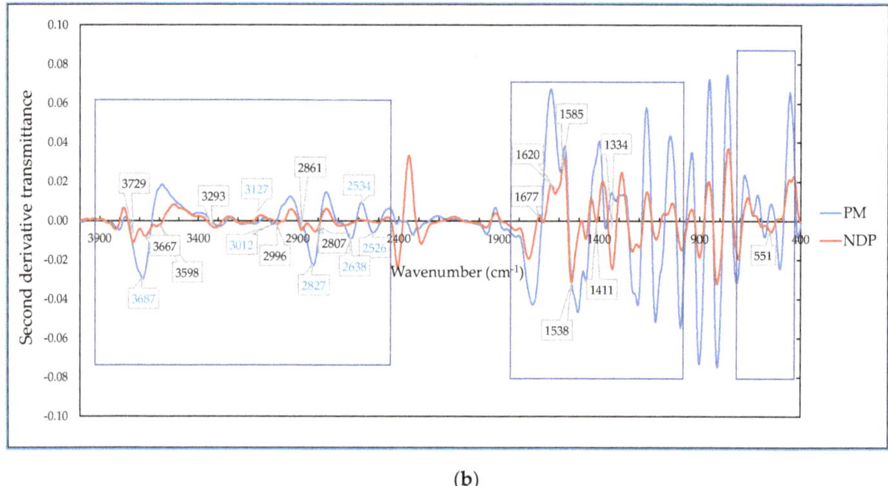

(b)

Figure 3. (**a**) FTIR spectra of diclofenac sodium hydrate (NDH), L-proline (LP), physical mixture 1:1 (PM), and the multicomponent 1:1 (NDP) from ethanol, methanol, and acetone. All NDPs have similar spectra, therefore only one of NDP spectra is marked on value of wavenumber changes. (**b**) The 2nd derivative FTIR spectra of NDP and PM. The blue number belongs to PM, and the black number reflects the NDP bands.

Figure 3b shows the broadband in the region of NDP at 2800–3700 cm^{-1}, which was known as the position of –OH stretching from water and carboxylic groups [26,39]. The -OH water molecule bands of NDH have been previously reported to exist on 3200–3500 cm^{-1}, coordinated with Na$^+$, and interacted by the hydrogen bond to the other molecules [26]. The functional group's broadband at 3200–3750 cm^{-1} region also may involve –NH and NH–O stretching. Meanwhile, 1600-1750 cm^{-1} included –NH bend and –C=O stretching of carboxylic from both starting materials of the intermolecular bond [21,26,38–46].

In more detail, derivative spectra of PM and cocrystal in Figure 3b show that the new bands at 3598, 3667, and 3729 cm^{-1} of NDP replaced the 3687 cm^{-1} band of PM, which indicated the change in OH stretching. Next, here were new NDP bands at 2807 and 2861 cm^{-1} and the disappearing of 2827, 2526, and 2638 cm^{-1} bands from PM. These data revealed the change of –OH stretching of carboxylic of the two materials due to the cocrystal formation. In addition, the distinctive bands at 1620 cm^{-1} and 1677 cm^{-1} on the NDP spectra also support the new bonding formation.

The other changes revealed were the shifting of the C=O band at 1573 cm^{-1} to 1585 cm^{-1} and the amino stretching band from 1400 cm^{-1} to 1411 cm^{-1}. This data reflected the alteration of COO$^-$ vibration of NDH and amino sidechains of LP [21,37,38]. A sharp band on 1334 cm^{-1} also supported the data of new interaction that involved –OH and C–O carboxylic. Besides, the changes of the 500–600 cm^{-1} band confirmed that –NH and CO moiety of LP contributed to composing the cocrystal. Therefore, overall, FTIR data supported a new solid structure construction, which was arranged by the carboxylate and amine from both starting materials, Na$^+$ ions from ND, and water molecules.

3.3.3. Thermal Analysis

Next, the electrothermal measurement on NDH showed a water spot at 70–100 °C, decomposed from 239.5 °C (change to a yellow mass), and finally melted at 288 °C. LP melted at 227 °C, as well as NDH, was continued by decomposition. Meanwhile, the new crystals released globules of water from about 50 °C, became very wet at ~86 °C, melted at ~116 °C as a clear liquid, and decomposed after ~200 °C, indicated by the color change and gas release. Thus, it can be concluded that the new cocrystal had a lower melting point than the starting components. In addition, all samples decomposed after melting.

The DSC and DTA/TG measurement results are presented in Figure 4. Figure 4a depicts a DSC thermogram overlay of NDH, LP, and NDP crystals. Figure 4a shows that NDH had endothermic curves at 50 and 70 °C, which indicates water release, next it melted at ~280 and oxidized at 288 °C. This data met the reference of ND tetrahydrate, as well as the FTIR spectra [26]. Meanwhile, LP melted and decomposed at 227 °C. Next, all thermograms of NDP cocrystals from aqueous 90% of methanol, ethanol, and acetone in this figure reveal the similar first endothermic peaks at 86.4, 85.2, and 85.3 °C; and the second peaks at 114.5, 115.2, and 116.5 °C, respectively. The first curves are predicted as the first step of water release, and the second curves indicate the last part of the water molecule release, which concurrently occurs with the melting point. It can be seen here that the melting point of NDP at 114–116 °C was significantly lower than the starting components (288 and 227 °C). The samples decomposed after 200 °C, which was indicated by an endothermic curve of DSC/DTA thermogram and the sharp loss of mass, as shown by the TG thermogram. These data in line electrothermal analysis results. The decrease of melting point reflects hydrogen bond formation between the compounds with the water molecule, which contributed crucially. The new thermal profile can be expected to impact on other physicochemical properties such as solubility, dissolution rate, and stability [14,15,20,47–51].

The DSC thermogram was then elaborated with DTA/TG data in Figure 4b, which shows that the mass decrease of NDP cocrystal was ± 14.2% *w/w* or equal to four water molecules. Hereafter, this new phase was estimated as a tetrahydrate form, named diclofenac sodium proline tetrahydrate (NDPT). A small difference in the transition temperature was shown between DSC and DTA/TG. The water release and melting point temperature on DTA/TG thermograms are shown similar to electrothermal data, but the endothermic peak of DTA was depicted lower than DSC. This difference was caused by the open pan used in DTA/TG versus the closed pan used in the DSC measurement.

The open pan used in DTA/TG measurement facilitated the loss of water freely, as did the open capillary tube used in the electrothermal study. Therefore, the endothermic peak of the melting point on the DSC thermogram was at the end of the DTA endothermic peak and the transition point of TG thermograms, 116 °C. Following the specification and the standard operation of the procedure analysis, the closed pan was preferred to maintain the DSC's instrument durability. DSC is more sensitive than DTA and highly influenced by liquid or gas contaminants. Prior analysis using electrothermal, which was elaborated with DTA/TG, showed that ND, LP, and the NDP cocrystal broke down and released water and gas; therefore, the closed pan was the best choice for DSC analysis.

The relationship between hydrate release and cocrystal melting mechanism of NDP can be studied from the DTA (red) and TG (blue) thermogram. The TG thermogram in Figure 4b shows that 10.7% mass, equal to three water molecules, decreased sharply in the temperature range of 50 to 85 °C. This following by a flatter curve until 100 °C, which shows the weight decreased up to 12.3% *w/w*, or equal to 3.5 water molecules. Finally, the last step of loss water, which demonstrated on the thermogram as −2% of weight, equivalent to half moles of water, was evident at 100–116 °C. This final water release was related to the endothermic peak at 107 °C on the DTA thermogram, which concurrently indicated the melting point of the cocrystal. These thermal data led to the prediction that the water molecules interacted in the different strengths in NDPT, and the last part of the water release was responsible for the cocrystal breaking.

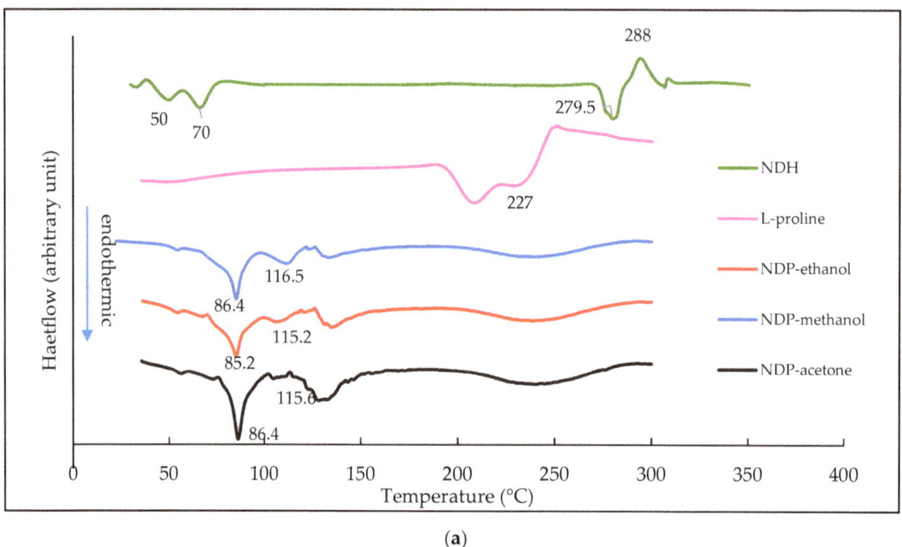

(**a**)

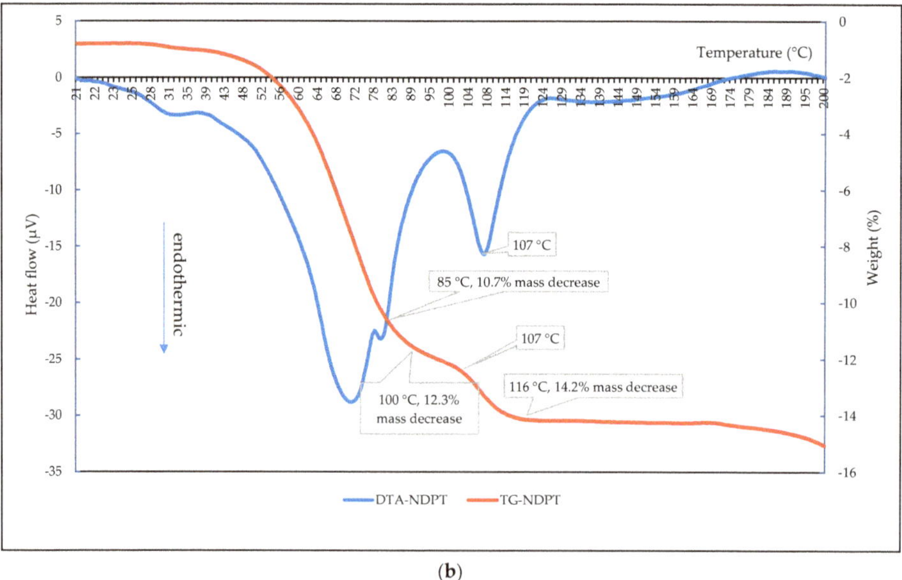

(**b**)

Figure 4. (**a**) Differential scanning calorimetry (DSC) thermogram of diclofenac sodium hydrate (NDH); L-proline (LP); and NDP cocrystals from acetone, ethanol, and methanol, respectively. (**b**) DTA and TG thermogram of NDP, shown ~14.2% mass decrease or equal to 4 water molecules.

3.3.4. PXRD Data

Figure 5 depicts the diffractograms of NDH, LP, and NDPT cocrystals from three kinds of solvent. NDH is shown in Figure 5a to have the specific peaks at $2\theta = 13.1°$, $13.8°$, $15.0°$, $22.2°$, and $26.5°$ (assigned by green marks). It was supported by the previous FTIR and TG data confirmed as a tetrahydrate, with water mass ~20% [26]. Next, it is shown in Figure 5b that the LP diffractogram had specific peaks at $2\theta = 15.2°$, $18.0°$, $19.5°$, $24.8°$ (purple marks). The diffractograms NDPT from all solvents that are shown in Figure 5c–e were similar, with the distinctive peaks are demonstrated at

2θ = 4.3°, 7.2°, 10.4°, 13.0°, 14.2°, 17.9°, 20.1°, 24.1°, and 26.3° (yellow marked). These data confirmed that a new multicomponent phase was produced from NDH and LP.

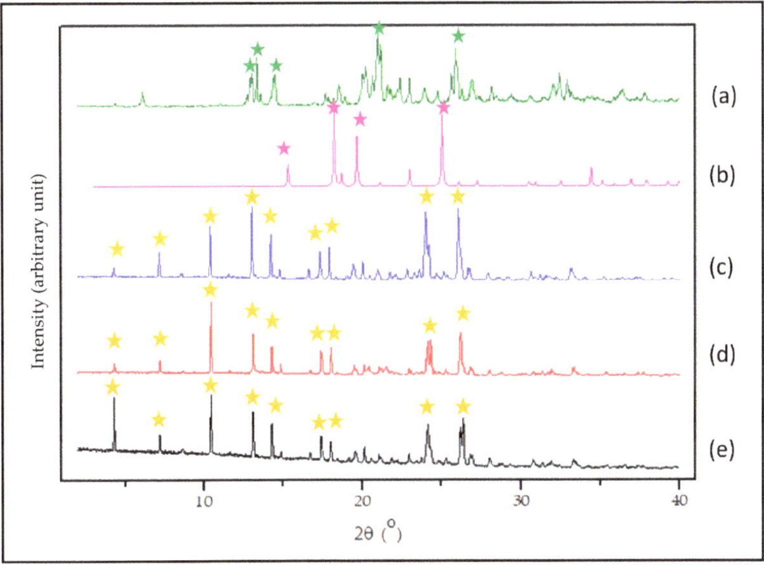

Figure 5. Diffractogram of NDH (**a**); LP (**b**); and NDPT from the 95% of methanol (**c**), acetone (**d**), and ethanol (**e**). The distinctive peaks of the new phase are assigned with yellow mark.

3.4. NDPT Crystal Structure Determination

The crystal structure of NDPT was determined by single-crystal X-ray analysis entirely, resulting in crystallographic data, which are listed in Table 1. From the study, NDP multicomponent crystal was confirmed to contain equimolar amounts of ND and LP. Furthermore, this crystal was found to be a hydrate with four water molecules for one molecule of ND and LP. Thus, it met with the TG result and was fixed to be a tetrahydrate. The salt cocrystal structure is depicted in Figure 6a, with the diffractogram is shown in Figure 6b.

Table 1. Crystallographic data of NDPT.

Crystal name	Diclofenac-Sodium-Proline-Tetrahydrate (NDPT)
Moiety formula	$C_{19}H_{27}Cl_2N_2NaO_8$, $(C_{14}H_{10}Cl_2NO_2Na)$ $(C_5H_9NO_2)$ $4(H_2O)$
Crystal system	Monoclinic
Space group	$P2_1$
a (+)	7.6073(3)
b (Å)	15.2652(7)
c (Å)	20.5263(9)
B (°)	97.602(1)
V (Å3)	2362.71(18)
Z/Z′	4/2
T (K)	296
R-factor (%)	5.93

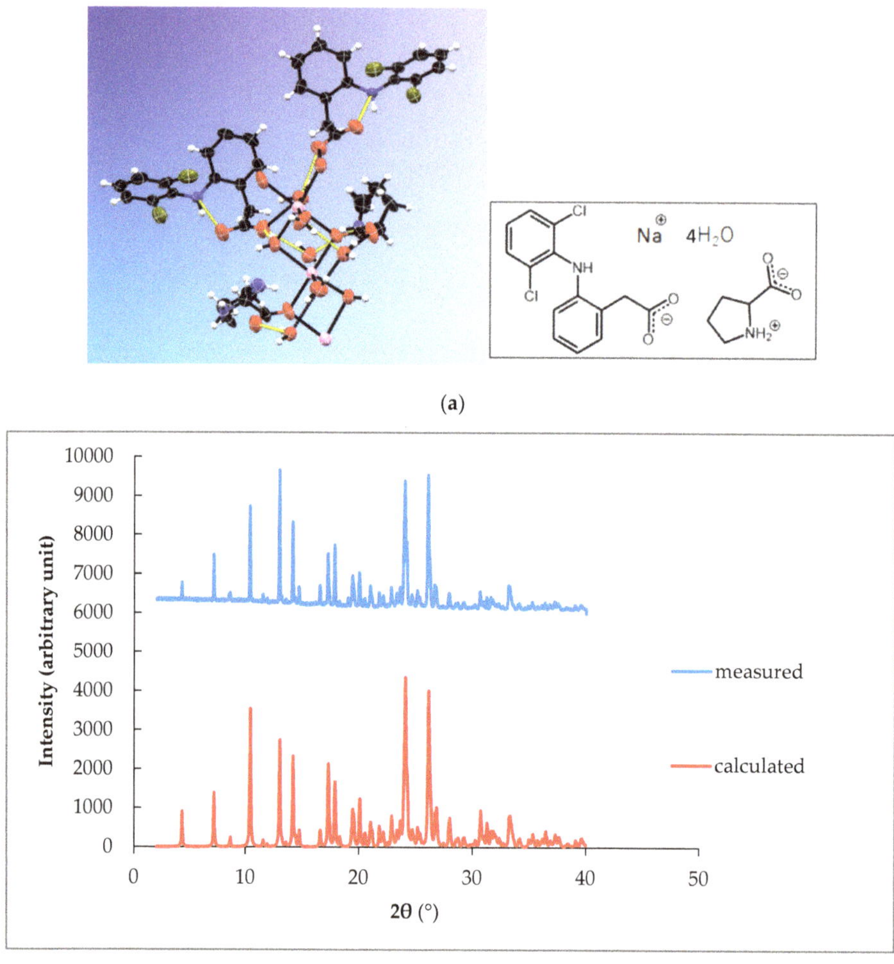

(a)

(b)

Figure 6. (**a**) ORTEP cell view and molecular structure graph of NDPT; Na$^+$ is in violet. (**b**) The measured and calculated diffractograms of NDPT.

The crystallographic data of NDPT are listed in Table 1, which revealed that this new phase is a monoclinic crystal with the space group $P2_1$.

Considering the atomic distance in the NDPT crystal structure, as shown in Figure 7, the C–O distance indicated that the diclofenac molecule was ionized. Based on this analysis, it can be concluded that the new crystal phase was a salt cocrystal, arranged from an ionized drug (diclofenac sodium) with a neutral coformer (LP) and supported by the water molecules as the bridge between the components. This interaction was different from the previous cocrystal of neutral diclofenac acid with LP [20]. The alkaline and ionized diclofenac can be expected to increase solubility and dissolution more than the neutral cocrystal, which then was proven further.

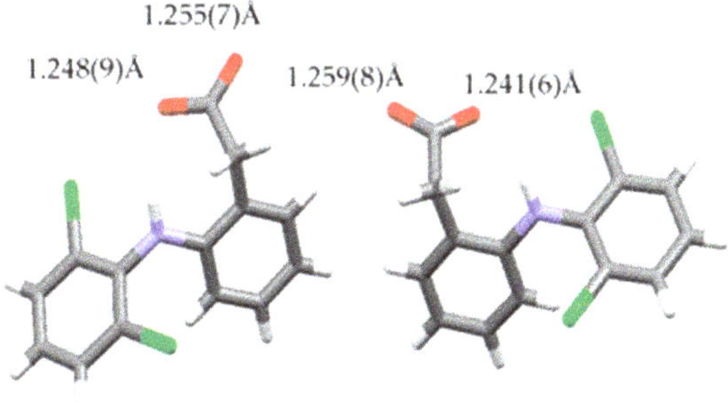

Figure 7. C–O distances in diclofenac molecules in NDPT.

Figure 8 presents the unique conformation of each component in the tetrahydrate salt cocrystals, which involves sodium, diclofenac, proline, and water. In the asymmetric unit of this crystal, there are two ND and two LP molecules and eight water molecules. From these water molecules, six molecules coordinate to Na$^+$, and two molecules form hydrogen bonds with other water molecules, LP molecules, and diclofenac molecules without coordination to Na$^+$. In addition, two LP molecules were also coordinated to Na$^+$, which was a 6-coordinated structure totally. These coordination bonds formed a one-dimensional (along a-axis) chain consisting of Na atom and water molecules, and both diclofenac molecules were connected to this Na chain by hydrogen bonds with water molecules. Furthermore, the carboxy oxygen of the diclofenac molecule formed a hydrogen bond with the nitrogen atom of LP. The formation of this hydrogen bond (C=O ⋯ HN) was also detected by an FTIR band at 1620 cm^{-1} and the change of 500–600 cm^{-1} band. The LP molecule was disordered at the 5-membered ring.

3.5. Diclofenac Sodium-Proline Monohydrate Salt Cocrystal Isolation, Characterization, and Structure Determination

We attempted to discover the other hydrate form of NDP by arranging the different circumstances of cocrystal production, called condition II. This condition was performed by storing all starting materials in a desiccator with silica gel for 24 h (0% RH/25 ± 2 °C) to minimize water existence. FTIR, TG, and PXRD were used to evaluate the solid state of ND and LP. As a result, we observed that even NDH back to ND, in line with the works in [25,26]. Based on the previous data of NDPT, the water molecule was shown to be prominent in the intermolecular interaction. From a screening, the starting materials of ND and LP were still found when the pure solvent (≥98%) was used. Thus, next cocrystal was collected using the purer solvents than that was used to produce NDPT, but it should still provide a small amount of water, i.e., the aqueous 90–95% *v/v* organic solvents. From this condition, a new crystal habit differed to NDPT was obtained successfully. The photographs present the newest crystalline morphology in Figure 9 (left) compared to the previous structure, NDPT, in Figure 9 (right).

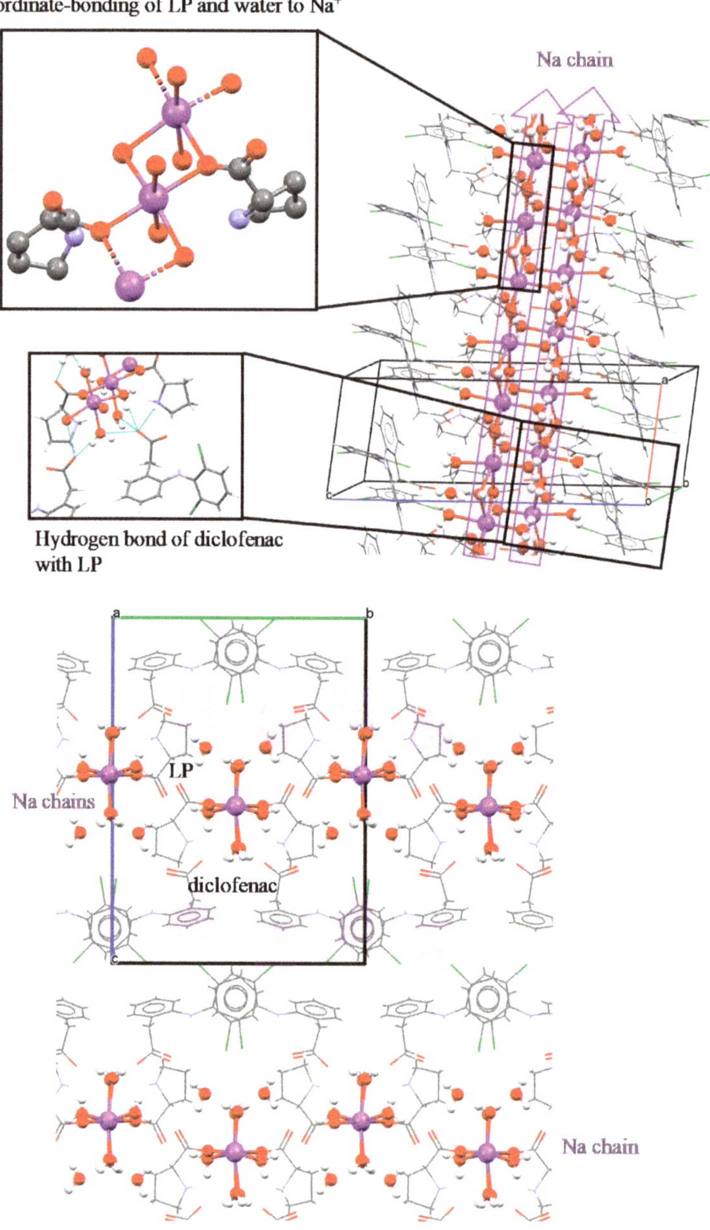

Figure 8. The Na chain structure in NDPT (diclofenac–sodium–proline–tetrahydrate), Na+ is in violet.

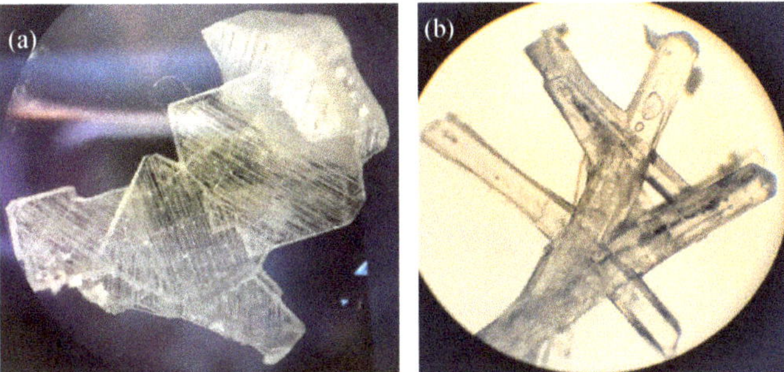

Figure 9. Crystal of the new diclofenac–sodium–proline–monohydrate (**a**) compared to NDPT (**b**) under a binocular microscope observation (100×).

The DTA/TG results of the newest cocrystal are revealed in Figure 10 below. Red and blue curves represent DTA and TG thermograms, respectively, which show the water release from room temperature to 200 °C. Approximately 2% of water released gradually at 30–100 °C, followed by a rapid loss of the left of ~ 2% water molecule (equal to 0.5 moles of water) at 100–116 °C, which occurred together with the melting point shown by a peak on DTA thermogram at 106 °C. It means that NDPM's melting point was not significantly different from NDPT. The total water release was ~4% *w/w*, or equal to a monohydrate. Therefore, it was named diclofenac sodium–proline–monohydrate (NDPM). Based on TG data, as well as shown in the NDPT thermogram, half moles of water molecule interacted strongly with the other components in NDP, which released lastly by breaking down the cocrystal completely.

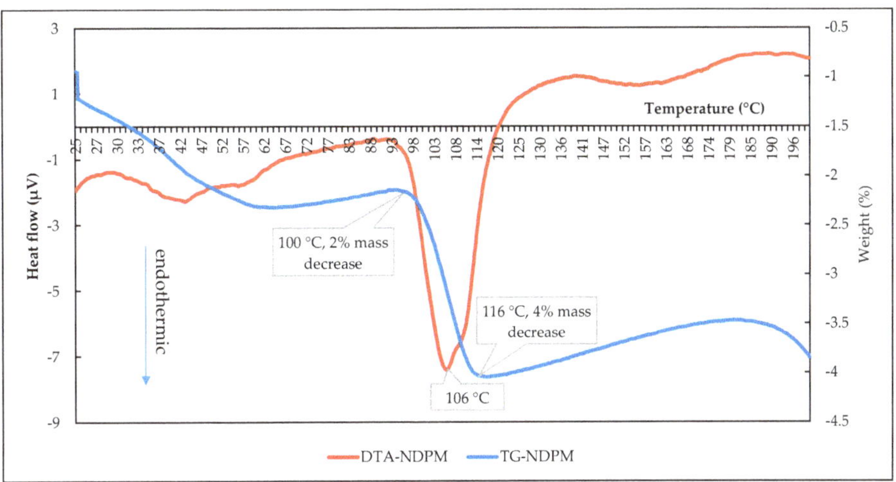

Figure 10. DTA and TG thermograms of NDPM.

PXRD analysis collected the diffractogram pattern of NDPM, which is depicted in Figure 11, showing the distinguished peaks at 2θ = 4.0°, 8.2°, 12.4°, and 13.3°; compared to NDPT peaks at 2θ = 4.3°, 7.2°, 10.4°, and 13.0°. This data confirmed that the new crystal obtained from the condition II was a different phase than NDPT, which then was determined further by SCXRD.

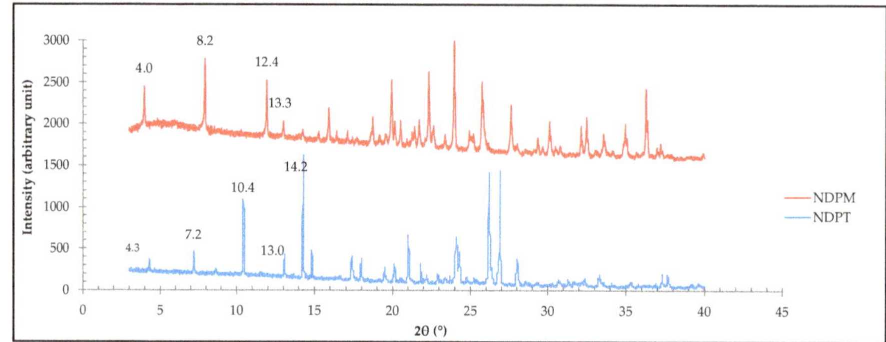

Figure 11. The measured diffraction pattern of NDPM (red) shows the different profile compared to NDPT (blue).

The structure determination using SCXRD at −180 °C confirmed that the new phase was a monohydrate, which consisted of diclofenac–sodium–proline–water (1:1:1:1). The ORTEP draw of NDPM structure is depicted in Figure 12a. In addition, Figure 12b reveals the similarity of the empirical and calculated diffractogram. Thus, the data show that NDPM structure has been determined entirely.

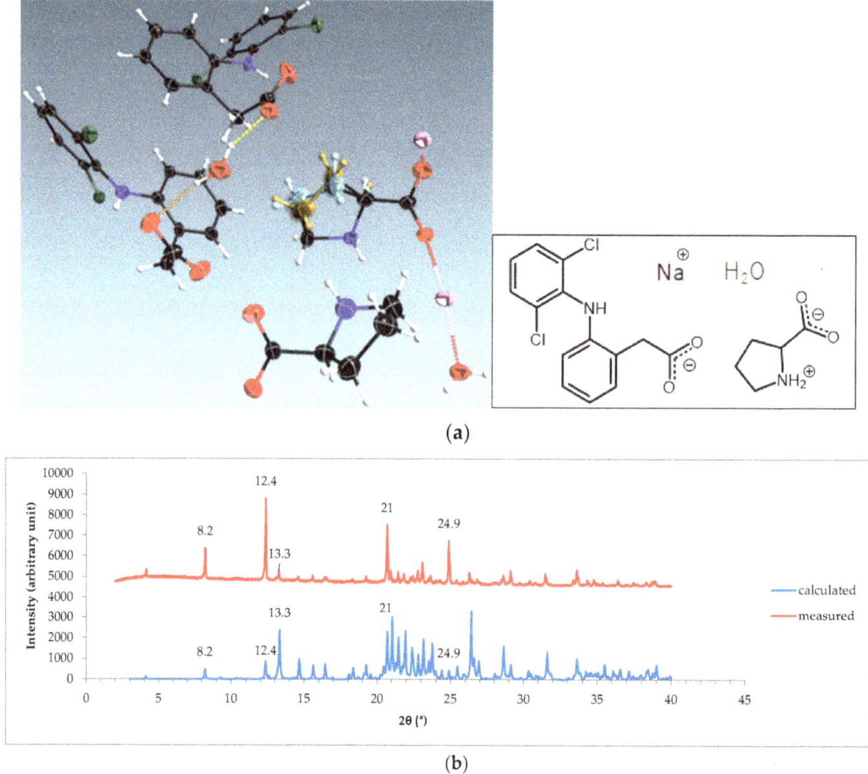

(a)

(b)

Figure 12. (**a**) ORTEP cell view and molecular structure graph of NDPM; Na^+ is in violet. (**b**) The measured and calculated diffractograms of NDPM.

The detailed element configuration of NDPM is shown in Figure 13. This figure presents that diclofenac is coordinated around the Na$^+$ network to form a layer structure. Meanwhile, LP allowed Na$^+$ to build a coordination network. However, the data of NDPM were not as perfect as the NDPT data, due to the instability and very thin crystal habit of this phase. Crystal NDPM was shown to be unstable and immediately transformed into NDPT, indicated by the habit change of the cocrystal under ambient conditions. This property was confirmed by TG/DTA and PXRD, which will be discussed further in Section 3.6 (Stability Testing).

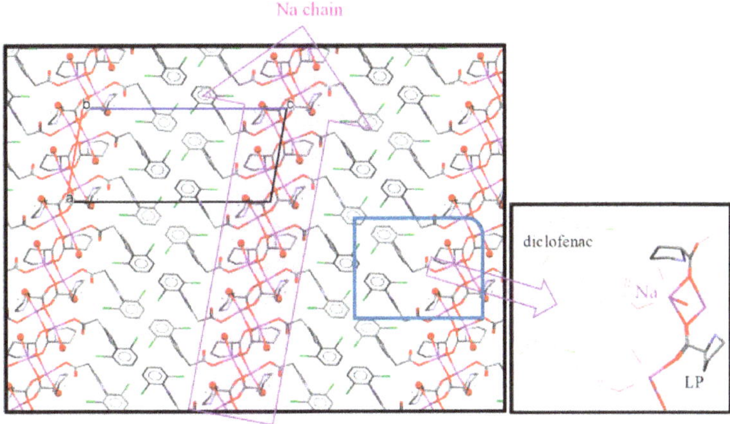

Figure 13. The order of diclofenac–sodium–proline (NDP) system. L-proline (LP) allows Na$^+$ to form a coordination work with diclofenac site. Na$^+$ is in violet.

The crystal data of NDPM is listed in Table 2 below.

Table 2. Crystallographic data of NDPM.

Page: 17 **Crystal Name**	Diclofenac-Sodium-Proline-Monohydrate (NDPM)
Crystal System	Monoclinic
Moiety formula	$C_{19}H_{21}Cl_2N_2NaO_5$, $(C_{14}H_{10}Cl_2NO_2Na)$ $(C_5H_9NO_2)$ (H_2O)
Space group	$P2_1$
a/Å	9.8240(3)
b/Å	9.2835(3)
c/Å	21.7931(6)
β/°	100.383(2)
V/Å^3	1955.01(10)
Z/Z′	4/2
R/%	7.43

In conclusion, the multicomponent of ND with LP was found in two forms, which were tetrahydrate (NDPT) and monohydrate (NDPM) salt cocrystals. The crystal data of NDPT and NDPM have been submitted to Cambridge Crystallographic Data Center (CCDC) with the deposit number 1938590 and 1983478, respectively. At the present time, the anhydrous phase is under investigation and not yet developed because it is difficult to compose, even in its powder form. This fact underlies the prediction that the presence of water is crucial in the formation of hydrogen bonds in this new salt cocrystal multicomponent system.

3.6. Stability Testing

Several stability studies are typically performed during the development of cocrystals, including drying and humidifying studies. The type of stability test depends on the structural characteristics of the sample [20,23,52–54]. Due to the four water molecules and the two LP molecules coordinated to Na^+ (sodium), it can be predicted that NDP can be transformed from one hydrate to the other phase easily, as well as ND [23–29]. The existence of a cationic element, mainly alkaline, has been well known to absorb water molecules quickly and release the water under dry conditions. Therefore, the hydrate stability test under drying and extreme humidity were essential to perform.

3.6.1. Thermal Profile Analysis on Dried Stability Test Samples

Dried samples were evaluated using DTA/TG/DSC and the data are shown in Figure 14. Figure 14a is DTA thermogram of the samples, show the decreasing of dehydrated curve at ~30–100 and 100–125 °C by the heating. Next, the TG thermograms in Figure 14b show the release of water of the hydrate at 2–6 h, leaving approximately one water molecule after 12 h or equal to a monohydrate salt cocrystal. It finally lost almost all water molecules after 24 h of drying in a controllable incubator, set at 30% RH/40 °C. DSC analysis was also conducted to confirm the thermal data of the new phases. The obtained thermograms are presented in Figure 14c, which are equal with the DTA/TG profile. These data were then elaborated with PXRD analysis, which will be discussed in Section 3.6.3.

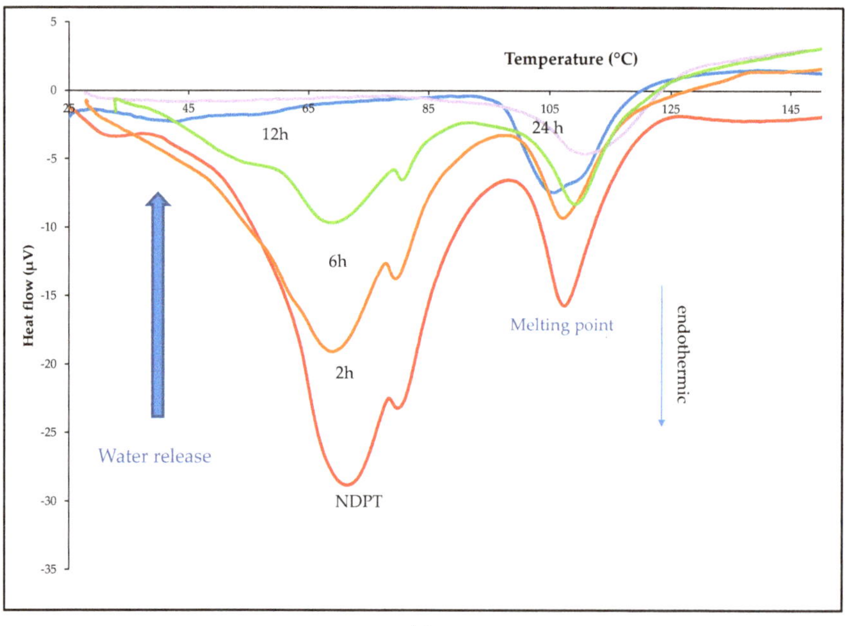

(a)

Figure 14. *Cont.*

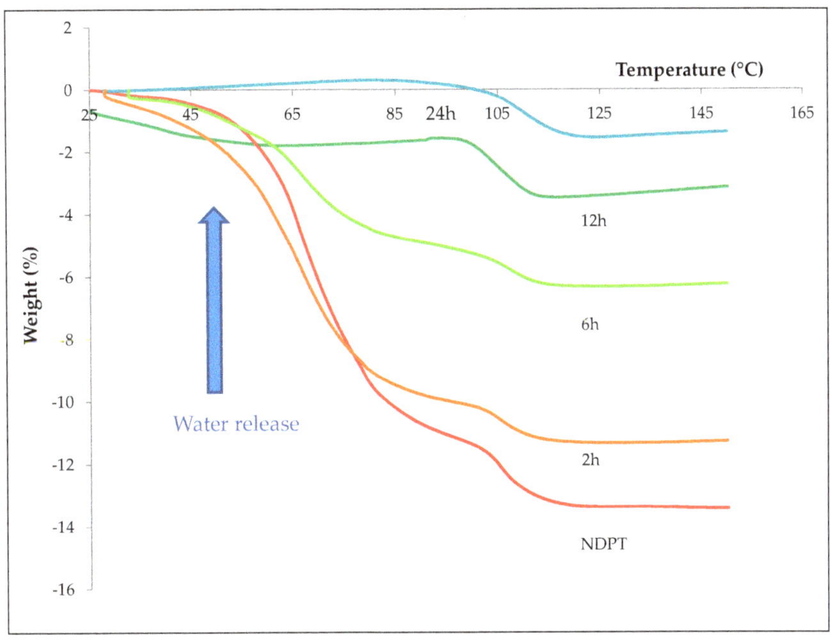

(b)

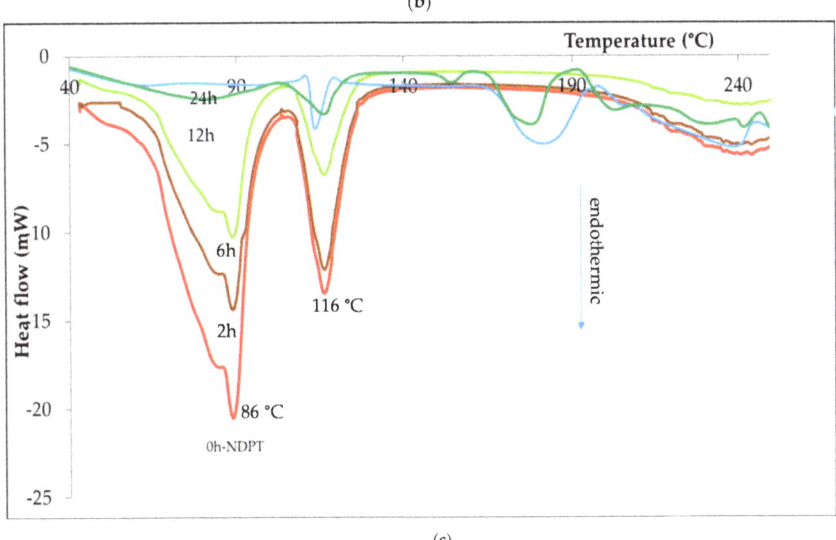

(c)

Figure 14. DTA (**a**), TG (**b**), and DSC (**c**) thermograms of NDPT sample that was stored in the incubator 30% RH/40 °C for 0, 2, 6, 12, and 24 h.

3.6.2. Thermal Profile Analysis on Humidified Stability Test Samples

Figure 15a shows the NDPT cocrystal thermograms from DSC after 24, 48, and 72 h of storage in a chamber of 94 ± 2% RH/25 ± 2 °C. The first endothermic peak at ~86 °C is a water release point and the second endothermic peak is shown at ± 116 °C. The slightly different profile of the 72 h and

15 days samples were shown in the first endothermic curve, separated into two small peaks. However, TG thermograms in Figure 15b reveal that the water portion did not increase significantly after 15 days, which was ~0.5% *w/w* or equal to 0.1 moles of water molecules only. This data indicated that the tetrahydrate was still stable. PXRD analysis confirmed NDPT stability under this humid condition, as explained in Section 3.6.3, concurrently with the drying stability test result.

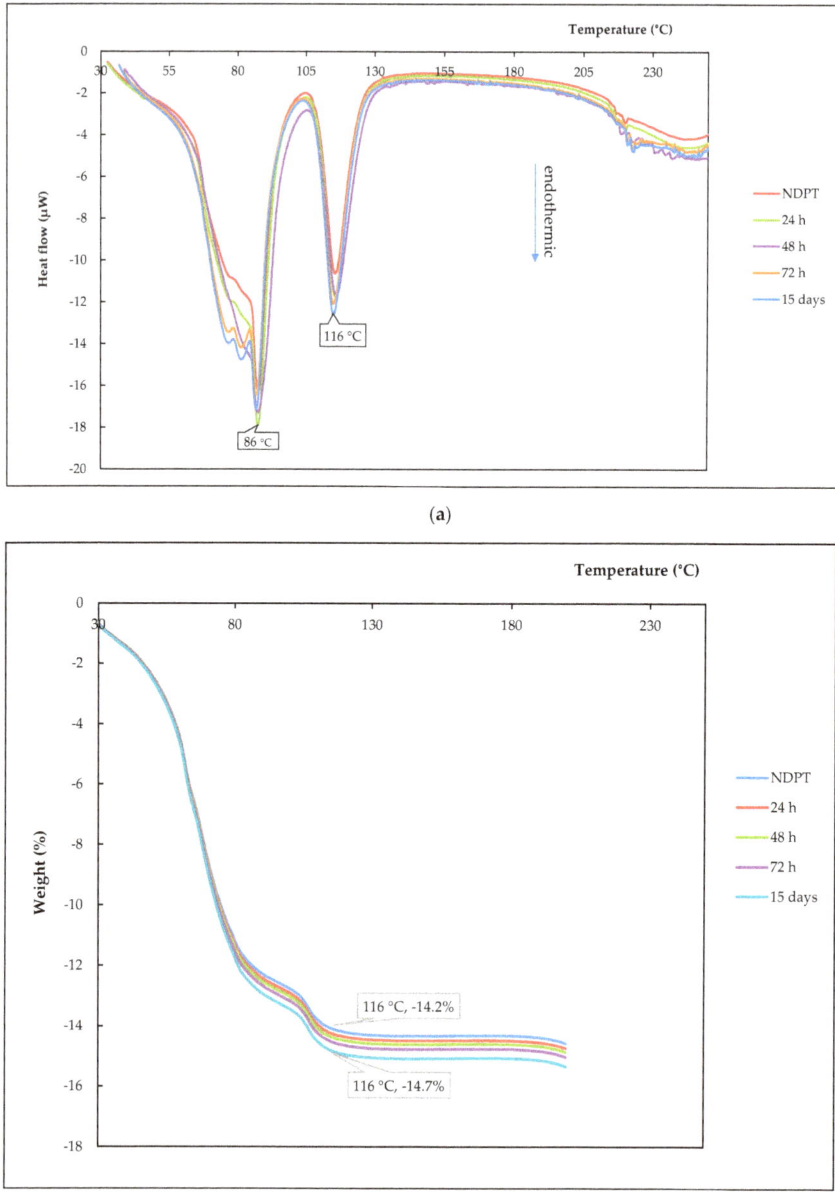

Figure 15. DSC (**a**) and TG (**b**) thermograms of NDPT after storage at high relative humidity of 94% RH/25 ± 2 °C for 24, 48, 72 h, and 15 days.

3.6.3. PXRD Analysis of Stability Test Samples after Drying and Humidifying

PXRD was then used to confirm the thermal analysis data from the stability test on NDPT samples, which were collected both by drying and storage in humid conditions, shown in Figure 16. ND, NDH, and LP diffractograms were depicted as the reference. Previously, it was predicted that the drying of NDPT would produce NDPM, as did by heating NDH to produce ND [25,26]; however, the diffractogram data revealed that the dried sample after 6 h of storage at 30 ± 0.5% RH/40 ± 0.5 °C showed a mixture peaks of NDPT and the starting materials, i.e., ND and LP, without any less hydrate of NDP. In this figure, the specific peaks of samples are assigned by the ticks in yellow (NDPT), blue (NDH), and green (LP). Moreover, NDPT changed to a similar diffraction pattern with a mixture of the components, anhydrous ND and LP, which are depicted in the figure as the comparison, after 12–24 h of drying. The ND peaks at 2θ = 6.7°, 8.6°, 15.2°, and 17.2° (blue ticks) appeared in the dried NDPT samples, as well as LP peaks at 2θ = 15.2°, 18.0°, 19.5°, and 24.8° (purple ticks). The final mixture did not contain NDH due to the dry condition desorbed all water molecules. These data describe that water loss broke cocrystal binding and did not result in another new phase, including NDPM. Thus, the water molecule was crucial in the NDPT system to mediate the interaction between components in this tetrahydrate salt cocrystal.

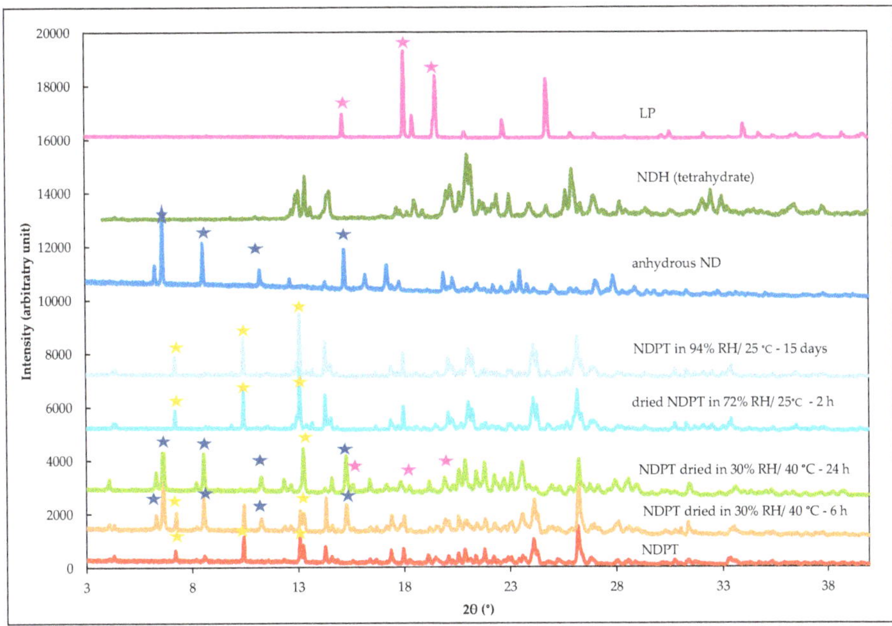

Figure 16. Diffractogram of NDPT fresh preparation after drying in 30% RH/40 °C for 6 h and 24 h, dried NDPT after 2 h restored in 72 ± 2% RH/25 ± 2 °C, and dried NDPT after restored in a 94% RH/25 ± 2 °C chamber for 15 days. Anhydrous ND, NDH (tetrahydrate), and LP diffractograms were depicted as the comparison. Assigned peaks represent ND (blue ticks), LP (purple ticks), and NDPT (yellow ticks).

Inversely, Figure 16 also shows that after 2 h restored under ambient conditions (72 ± 2% RH/25 ± 2 °C), the diffractogram of the sample was back to the NDPT profile, with the specific peaks at 2θ = 4.3°, 7.2°, 13.0° (yellow ticks), without producing NDPM. Moreover, NDPT was still stable under high humidity, i.e., 94 ± 2% RH/25 ± 2 °C, until 15 days, indicated by the steady diffractograms. In conclusion, NDPT was a stable phase, comparable to NDH pseudo polymorphs' stability, which has been reported to be more stable than ND [25,26,28,29]. Anhydrous ND changed rapidly under ambient conditions to its hydrate (NDH) [25,26], similar to NDPM, which quickly transformed into NDPT. However, it differs from NDH-ND transformation, which can occur easily by drying under the dry condition (30% RH/40 °C), NDPT could not transform to the less hydrate by this manner.

3.7. Pharmaceutics Performance Tests

The pharmaceutical performance of the new salt cocrystals was carried out by investigating the impact of the multicomponent formation on ND solubility and the dissolution profile. The tests were performed with NDPM and NDPT, compared to anhydrous ND.

3.7.1. Solubility Testing

Solubility is of utmost importance in the pharmaceutical field because the drug must dissolve before being absorbed. Table 3 depicts the solubility data for ND, NDPM, and NDPT. The solubility of the physical mixture of anhydrous ND with LP was conducted to represent the dried system, which was not found as the monohydrate cocrystal. Based on the solubility test results, after the equilibrium state, diclofenac was dissolved more readily from NDPM and NDPT, i.e., 4.08 and 3.31-fold increase compared ND solubility, respectively. NDPM was revealed to have higher solubility than NDPT, comparable to anhydrous ND behavior, which was superior to ND hydrate due to the lower of crystal lattice energy with a wider space between molecules [24–29]. The increased solubility of the NDP systems was related to the decrease in lattice energy, which promoted the breaking down of solid intermolecular bonds immediately [20,46,48–51,54].

In addition, the characteristic crystal structure associated with solubility in the NDP salt cocrystals was in the presence of a region consisting only of Na^+ ions and water molecules (Na chain) that has a sufficiently higher affinity for water than diclofenac acid and LP. Because of this higher affinity, Na^+ ions and water molecules would be dissolved first through the chain region. As a result, the entire NDP crystal breaks up and dissolves (reduction in lattice energy stabilization). Therefore, the presence of the Na chain and the LP layer would be one of the reasons why the solubility was higher than that of ND. Several studies have reported that such a layer or channel structure consisting only of higher solubility molecules improves the solubility of API multicomponent crystals [50,51].

Table 3. Solubility curve (in CO_2-free distilled water, pH 7.0) of ND, NDPM, and NDPT.

Sample	Solubility (mg/mL) $n = 3$	SD	Solubility Magnitude of ND	pH of Solution
ND	16.18	0.46	1.00	6.9–7.0
NDPM	66.09	1.65	4.08	7.0–7.1
NDPT	53.61	0.83	3.31	7.0–7.1

3.7.2. Dissolution Testing

Dissolution testing was performed to observe how salt cocrystal formation affected the ND release rate. Tests were carried out with samples equivalent to 200 mg ND in 900 mL gastric medium (pH 1.2) and intestinal medium (pH 6.8) [25,26,48,55–57]. Diclofenac solubility has been reported to be influenced by pH medium and has the optimum dissolution in pH neutral to base [25,48,57]. Based on dissolution profiles on pH 1.2 medium (Figure 17a), the maximum drug release from all samples (ND, NDPM, and NDPT) was achieved after 15 min dissolution, but in a very low concentration. This result was in line with the information from references [26,49,58]. Due to the acidity of ND (pKa: 4, at 25 °C), it showed that in the gastric medium, each sample was not completely separated because hydrogen ions inhibited diclofenac dissociation. However, NDPM was superior in increasing the percentage of drug release. Based on the Henderson–Hasselbach's equation at a pH below the pKa of

ND, the percentage of non-ionized substance will be greater than at a pH above the pKa so that the percentage of substances absorbed into the solute tends to increase [55,57,58].

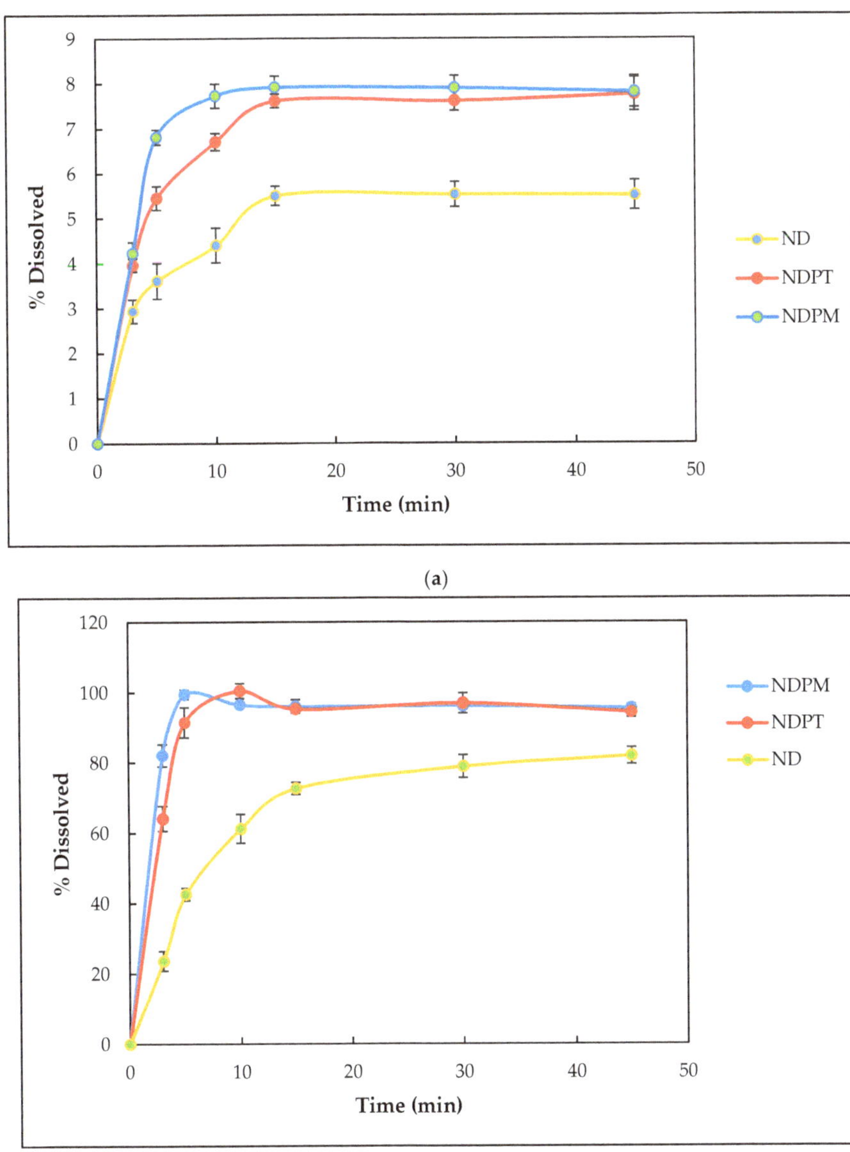

(a)

(b)

Figure 17. Dissolution profile of anhydrous diclofenac sodium (ND), salt cocrystal diclofenac–sodium–proline–monohydrate (NDPM), and salt cocrystal diclofenac–sodium–proline–tetrahydrate (NDPT) in (a) gastric pH (1.2) medium and (b) intestinal pH (6.8) medium.

ND dissolved in the pH 6.8 medium. However, it needed time, which was represented by the dissolution profile in Figure 17b. This figure shows that NDPM dissolution was fastest at intestinal pH, which reached a maximum concentration (100% released of 200 mg drug) after 5 min. NDPT needed only 10 min. Meanwhile, the ND curve still increased after 45 min. This means that NDPM was superior in releasing diclofenac in pH 6.8 medium compared to the single ND and the NDPT. As shown in this figure, the existence of LP can push the dissolution, even in the physical mixture. However, the cocrystal formation made the interaction between ND and LP intensively by a stronger hydrogen bonding than the physical mixture. The very immediate entire release of NDPT occurs because the hydroxyl ion can accelerate the dissociation of diclofenac.

The increase in diclofenac solubility from NDPM and NDPT was due to the ability of LP to arrange the coordinate binding with Na^+ and interact as a zwitterion molecule with diclofenac acid. This extraordinary double action of LP facilitated contact between the diclofenac moiety and the dissolution medium, in addition to the dissociation of Na^+ with diclofenac molecule in the media. By increasing dissolution, NDP salt cocrystals can be expected to improve diclofenac bioavailability [6,59]. The dissolution profile of the salt cocrystals was better than the previous non-salt cocrystal of diclofenac–proline [20], as the sink dissolution conditions can lead to problems with the absorption step of the drug [47,60] did not exist. ND had a good dissolution profile in the buffered pH 6.8 medium, but the cocrystals were superior. The amphoteric ability of LP may adjust the pH to the optimum environment for diclofenac dissolution (the medium pH was measured as 6.8–7.0 after the final dissolution), along with hydrotropic property, which contributes to improve dissolution. NDPM showed a faster dissolution than NDPT for a similar reason to the solubility data. The monohydrate crystal lattice is smaller and has a looser and more spacious space compared to the tetrahydrate form, as presented in Figures 8 and 14, making it dissolve faster. This data was in accordance with anhydrous ND, which has a higher solubility than its hydrate forms [26,28,29].

4. Conclusions

ND and LP form two pseudo-polymorphs of multicomponent salt cocrystals. The first multicomponent crystal involves four water molecules, or a tetrahydrate, named NDPT. Meanwhile, the second consists of one water molecule, or a monohydrate, named NDPM. Both hydrates have a monoclinic lattice system. However, the monohydrate form was unstable and changed to NDPT easily under ambient condition. NDPT was a stable salt cocrystal even after breaking the structure after drying due to water loss; its structure recovered rapidly under ambient conditions. Moreover, NDPT structure was also stable under high humidity. Both salt cocrystals significantly increased the solubility and dissolution rate of the original starting material, ND, but NDPM was superior. Further research should hold continuously to observe other polymorphs and hydrates, which is still very challenging. The investigation of diffusion and absorption should also be carried out.

Supplementary Materials: The following are available online at http://www.mdpi.com/1999-4923/12/7/690/s1: S1: CIF crystal data of NDP monohydrate, S2: CIF crystal data of NDP tetrahydrate, S3: Diffractogram (1) and TG-thermogram (2) of ND and NDH.

Author Contributions: Conceptualization, I.N.; methodology, I.N., R.A.K., H.U., and A.H.; software, R.A.K. and A.H.; validation, I.N., R.A.K., A.H., and H.U.; formal analysis, I.N., R.A.K., W.A., and A.H.; investigation, I.N., R.A.K., and A.H.; resources, I.N. and H.U.; data curation, I.N., A.H., and H.U.; writing—original draft preparation, I.N., R.A.K., and H.U.; writing—review and editing, I.N., A.H., H.U., and W.N.A.; visualization, I.N., A.H., and W.N.; supervision, I.N. and H.U.; project administration, I.N. and W.N.; funding acquisition, I.N. and H.U. All authors have read and agreed to the published version of the manuscript.

Funding: This research was funded by High Educational Directorate of Indonesia Education Ministry of Republic Indonesia—Research and Society Service Institution of Bandung Institute of Technology, Number 2/E1/KP.PTNBH/2020 (IN). A part of this study was performed in Tokyo Institute of Technology, Japan, and was supported by JSPS KAKENHI Grant Number JP17K05745 and JP18H04504 (HU).

Acknowledgments: This research was funded by High Educational Directorate of Indonesia Education Ministry of Republic Indonesia—Research and Society Service Institution of Bandung Institute of Technology, Number 2/E1/KP.PTNBH/2020 (IN). A part of this study was performed in Tokyo Institute of Technology, Japan, and was supported by JSPS KAKENHI Grant Number JP18H04504 and 20H04661 (HU).

Conflicts of Interest: The authors declare no conflicts of interest.

Abbreviations

ND	natrium diclofenac (=diclofenac sodium)
NDH	natrium diclofenac hydrate (=diclofenac sodium hydrate)
LP	L-proline
NDP	natrium diclofenac proline
NDPT	natrium diclofenac proline tetrahydrate
NDPM	natrium diclofenac proline monohydrate
DSC	differential scanning calorimetry
DTA	differential thermal analysis
TG	thermogravimetry
FTIR	Fourier transform infrared
PXRD	powder X-ray diffractometer/powder X-ray diffractometry
SCXRD	single crystal X-ray diffractometry/single crystal X-ray diffractometer
RH	relative humidity

References

1. Arisoy, G.G.; Dural, E.; Mergen, G.; Arisoy, M.; Guvendik, G.; Soylemezoglu, T. Development and validation HPLC-UV method for the determination of diclofenac in human plasma with application to pharmacokinetic study. *Turk. J. Pharm. Sci.* **2016**, *13*, 292–299.
2. McGettigan, P.; Henry, D. Use of non-steroidal anti-inflammatory drugs that elevate cardiovascular risk: An examination of sales and essential medicines lists in low-, middle-, and high-income countries. *PLoS Med.* **2013**, *10*, e1001388. [CrossRef] [PubMed]
3. Vemula, V.R.; Lagishetty, V.; Lingala, S. Solubility enhancement techniques. *Int. J. Pharm. Sci. Rev. Res.* **2010**, *5*, 41–51.
4. Savjani, K.T.; Gajjar, A.K.; Savjani, J.K. Drug solubility: Importance and enhancement techniques. *ISRN Pharm.* **2012**, *2012*, 195727. [CrossRef] [PubMed]
5. Soares, F.L.F.; Carneiro, R.L. Evaluation of analytical tools and multivariate methods for quantification of co-former crystals in ibuprofen-nicotinamide co-crystals. *J. Pharm. Biomed. Anal.* **2014**, *89*, 166–175. [CrossRef] [PubMed]
6. Tilborg, A.; Springuel, G.; Norberg, B.; Wouters, J.; Leyssens, T. On the influence of using a zwitterionic coformer for cocrystallization: Structural focus on naproxen–proline cocrystals. *CrystEngComm* **2013**, *15*, 3341. [CrossRef]
7. Sathisaran, I.; Dalvi, S. Engineering cocrystals of poorly water-soluble drugs to enhance dissolution in aqueous medium. *Pharmaceutics* **2018**, *10*, 108. [CrossRef]
8. Fukte, S.; Wagh, M.P.; Rawat, S. Coformer selection: An important tool in cocrystal formation. *Int. J. Pharm. Pharm. Sci.* **2014**, *6*, 9–14.
9. Kuminek, G.; Cao, F.; Bahia de Oliveira da Rocha, A.; Gonçalves Cardoso, S.; Rodríguez-Hornedo, N. Cocrystals to facilitate delivery of poorly soluble compounds beyond-rule-of-5. *Adv. Drug Deliv. Rev.* **2016**, *101*, 143–166. [CrossRef]
10. Aitipamula, S.; Banerjee, R.; Bansal, A.K.; Biradha, K.; Cheney, M.L.; Choudhury, A.R.; Desiraju, G.R.; Dikundwar, A.G.; Dubey, R.; Duggirala, N.; et al. Polymorphs, salts, and cocrystals: What's in a name? *Cryst. Growth Des.* **2012**, *12*, 2147–2152. [CrossRef]
11. Qiao, N.; Li, M.; Schlindwein, W.; Malek, N.; Davies, A.; Trappitt, G. Pharmaceutical cocrystals: An overview. *Int. J. Pharm.* **2011**, *419*, 1–11. [CrossRef] [PubMed]
12. Patole, T.; Deshpande, A. Co-crystallization-a technique for solubility enhacement. *Int. J. Pharm. Sci. Res.* **2014**, *5*, 3566–3576.
13. Barikah, K.Z. Traditional and novel methods for cocrystal formation: A mini review. *Syst. Rev. Pharm.* **2018**, *9*, 79–82. [CrossRef]

14. Karimi-Jafari, M.; Padrela, L.; Walker, G.M.; Croker, D.M. Creating cocrystals: A review of pharmaceutical cocrystal preparation routes and applications. *Cryst. Growth Des.* **2018**, *18*, 6370–6387. [CrossRef]

15. Schultheiss, N.; Newman, A. Pharmaceutical cocrystals and their physicochemical properties. *Cryst. Growth Des.* **2009**, *9*, 2950–2967. [CrossRef]

16. Greenspan, L. Humidity fixed points of binary saturated aqueous solutions. *J. Res. Nat. Bur. Stand-A Phys. Chem.* **1977**, *81A*, 89–96. [CrossRef]

17. Grifasi, F.; Chierotti, M.R.; Gaglioti, K.; Gobetto, R.; Maini, L.; Braga, D.; Dichiarante, E.; Curzi, M. Using salt cocrystals to improve the solubility of niclosamide. *Cryst. Growth Des.* **2015**, *15*, 1939–1948. [CrossRef]

18. Gunnam, A.; Suresh, K.; Nangia, A. Salts and salt cocrystals of the antibacterial drug pefloxacin. *Cryst. Growth Des.* **2018**, *18*, 2824–2835. [CrossRef]

19. Nugrahani, I.; Pertiwi, E.A.; Putra, O.D. Theophylline-Na-saccharin single crystal isolation for its structure determination. *Int. J. Pharm. Sci.* **2015**, *7*, 15–24.

20. Nugrahani, I.; Utami, D.; Ibrahim, S.; Nugraha, Y.P.; Uekusa, H. Zwitterionic cocrystal of diclofenac and L-proline: Structure determination, solubility, kinetics of cocrystallization, and stability study. *Eur. J. Pharm. Sci.* **2018**, *117*, 168–176. [CrossRef] [PubMed]

21. Bucci, R.; Magri, A.D.; Magri, A.L. Determination of diclofenac salts in pharmaceutical formulations. *Fresenius' J. Anal. Chem.* **1998**, *362*, 577–582. [CrossRef]

22. Adeyeye, C.M.; Li, P.K. Diclofenac Sodium. In *Analytical Profiles of Drug Substances*; Florey, K., Ed.; Academic Press: London, UK, 1990; Volume 19, pp. 123–140.

23. Fini, A.; Cavallari, C.; Bassini, G.; Ospitali, F.; Morigi, R. Diclofenac salts, part 7: Are the pharmaceutical salts with aliphatic amines stable? *J. Pharm. Sci.* **2012**, *101*, 157–3168. [CrossRef] [PubMed]

24. Reck, G.; Faust, G.; Dietz, G. X-ray crystallographic studies of diclofenac-sodium-structural analysis of diclofenac-sodium tetrahydrate. *Pharmazie* **1988**, *43*, 771–774.

25. Fini, A.; Garuti, M.; Fazio, G.; Alvarez-Fuentes, J.; Holgado, M.A. Diclofenac salts. I. fractal and thermal analysis of sodium and potassium diclofenac salts. *J. Pharm. Sci.* **2001**, *90*, 2049–2057. [CrossRef]

26. Bartolomei, M.; Bertocchi, P.; Antoniella, E.; Rodomonte, A. Physico-chemical characterisation and intrinsic dissolution studies of a new hydrate form of diclofenac sodium: Comparison with anhydrous form. *J. Pharm. Biomed. Anal.* **2006**, *40*, 1105–1113. [CrossRef]

27. Muangsin, N.; Prajaubsook, M.; Chaichit, N.; Siritaedmukul, K.; Hannongbua, S. Crystal structure of a unique sodium distorted linkage in diclofenac sodium pentahydrate. *Anal. Sci.* **2002**, *18*, 967–968. [CrossRef] [PubMed]

28. Bartolomei, M.; Rodomonte, A.; Antoniella, E.; Minelli, G.; Bertocchi, P. Hydrate modifications of the non-steroidal anti-inflammatory drug diclofenac sodium: Solid-state characterization of a trihydrate form. *J. Pharm. Biomed. Anal.* **2007**, *45*, 443–449. [CrossRef]

29. Llinàs, A.; Burley, J.C.; Box, K.J.; Glen, R.C.; Goodman, J.M. Diclofenac solubility: Independent determination of the intrinsic solubility of three crystal forms. *J. Med. Chem.* **2007**, *50*, 979–983. [CrossRef]

30. Yang, Z. Natural Deep Eutectic Solvents and Their Applications in Biotechnology. In *Application of Ionic Liquids in Biotechnology*; Springer: Cham, Switzerland, 2018.

31. L-Proline. Available online: https://pubchem.ncbi.nlm.nih.gov/compound/L-proline (accessed on 17 July 2019).

32. Liu, Y.; Friesen, J.B.; McAlpine, J.B.; Lankin, D.C.; Chen, S.-N.; Pauli, G.F. Natural deep eutectic solvents: Properties, applications, and perspectives. *J. Nat. Prod.* **2018**, *81*, 679–690. [CrossRef]

33. Cherukuvada, S.; Guru Row, T.N. Comprehending the formation of eutectics and cocrystals in terms of design and their structural interrelationships. *Cryst. Growth. Des.* **2014**, *14*, 4187–4198. [CrossRef]

34. Yamashita, H.; Hirakura, Y.; Yuda, M.; Teramura, T.; Terada, K. Detection of cocrystal formation on binary phase diagrams using thermal analysis. *Pharm. Res.* **2013**, *30*, 70–80. [CrossRef]

35. Nugrahani, I.; Asyarie, S.; Soewandhi, S.N.; Ibrahim, S. The cold contact methods as a simple drug interaction detection system. *Res. Lett. Phys. Chem.* **2008**, *2008*, 169247. [CrossRef]

36. Karagianni, A.; Malamatari, M.; Kachrimanis, K. Pharmaceutical cocrystals: New solid phase modification approaches for the formulation of APIs. *Pharmaceutics* **2018**, *10*, 18. [CrossRef] [PubMed]

37. Rieppo, L.; Saarakkala, S.; Närhi, T.; Helminen, H.J.; Jurvelin, J.S.; Rieppo, J. Application of second derivative spectroscopy for increasing molecular specificity of fourier transform infrared spectroscopic imaging of articular cartilage. *Osteoarthr. Cartil.* **2012**, *20*, 451–459. [CrossRef] [PubMed]

38. Garbacz, P.; Wesolowski, M. DSC, FTIR and Raman Spectroscopy coupled with multivariate analysis in a study of co-crystals of pharmaceutical interest. *Molecules* **2018**, *23*, 2136. [CrossRef] [PubMed]

39. Szakonyi, G.; Zelkó, R. Water content determination of superdisintegrants by means of ATR-FTIR Spectroscopy. *J. Pharm. Biomed. Anal.* **2012**, *63*, 106–111. [CrossRef]

40. Jegatheesan, A.; Murugan, J.; Neelakantaprasad, B.; Rajarajan, G. FTIR, PXRD, SEM, RGA investigations of ammonium dihydrogen phosphate (ADP) single crystal. *Int. J. Comput. Appl.* **2012**, *53*, 15–18.

41. Ali, H.R.H.; Alhalaweh, A.; Mendes, N.F.C.; Ribeiro-Claro, P.; Velaga, S.P. Solid-state vibrational spectroscopic investigation of cocrystals and salt of indomethacin. *Cryst. Eng. Comm.* **2012**, *14*, 6665. [CrossRef]

42. Nanjwade, V.; Manvi, F.; Shamrez, A.; Nanjwade, B. Characterization of prulifloxacin-salicylic acid complex by IR, DSC and PXRD. *J. Pharm. Biomed. Sci.* **2011**, *5*, 1–6.

43. Brittain, H.G. Vibrational Spectroscopic study of the cocrystal products formed by cinchona alkaloids with 5-nitrobarbituric acid. *J. Spectrosc.* **2015**, *2015*, 340460. [CrossRef]

44. Ryu, I.S.; Liu, X.; Jin, Y.; Sun, J.; Lee, Y.J. Stoichiometrical analysis of competing intermolecular hydrogen bonds using infrared spectroscopy. *RSC Adv.* **2018**, *8*, 23481–23488. [CrossRef]

45. Zhang, G.C.; Lin, H.L.; Lin, S.Y. Thermal analysis and FTIR spectral curve-fitting investigation of formation mechanism and stability of indomethacin-saccharin cocrystals via solid-state grinding process. *J. Pharm. Biomed. Anal.* **2012**, *66*, 162–169. [CrossRef] [PubMed]

46. Cugovčan, M.; Jablan, J.; Lovrić, J.; Cinčić, D.; Galić, N.; Jug, M. Biopharmaceutical characterization of praziquantel cocrystals and cyclodextrin complexes prepared by grinding. *J. Pharm. Biomed. Anal.* **2017**, *137*, 42–53. [CrossRef]

47. Salas-Zúñiga, R.; Rodríguez-Ruiz, C.; Höpfl, H.; Morales-Rojas, H.; Sánchez-Guadarrama, O.; Rodríguez-Cuamatzi, P.; Herrera-Ruiz, D. Dissolution advantage of nitazoxanide cocrystals in the presence of cellulosic polymers. *Pharmaceutics* **2020**, *12*, 23. [CrossRef] [PubMed]

48. Nugrahani, I.; Utami, D.; Ayuningtyas, L.; Garmana, A.N.; Oktaviary, R. New preparation method using microwave, kinetics, in vitro dissolution-diffusion, and anti-inflammatory study of diclofenac-proline co-crystal. *ChemistrySelect* **2019**, *4*, 13396–13403. [CrossRef]

49. Roy, L.; Lipert, M.P.; Rodríguez-Hornedo, N. *Co-Crystal Solubility and Thermodynamic Stability, Pharmaceutical Salts and Co-Crystals*; RCS Publishing: Cambridge, UK, 2012; pp. 249–250.

50. Putra, O.D.; Umeda, D.; Nugraha, Y.P.; Furuishi, T.; Nagase, H.; Fukuzawa, K.; Uekusa, H.; Yonemochi, E. Solubility improvement of epalrestat by layered structure formation: Via cocrystallization. *Cryst. Eng. Comm.* **2017**, *19*, 2614–2622. [CrossRef]

51. Putra, D.O.; Umeda, D.; Fujita, E.; Haraguchi, T.; Uchida, T.; Yonemochi, E.; Uekusa, H. Solubility improvement of benexate through salt formation using artificial sweetener. *Pharmaceutics* **2018**, *10*, 64. [CrossRef]

52. Eyjolfsson, R. Diclofenac sodium: Oxidative degradation in solution and solid state. *Drug Dev. Ind. Pharm.* **2000**, *26*, 451–453. [CrossRef]

53. Eddleston, M.D.; Madusanka, N.; Jones, W. Cocrystal dissociation in the presence of water: A general approach for identifying stable cocrystal forms. *J. Pharm. Sci.* **2014**, *103*, 2865–2870. [CrossRef]

54. Prenner, E.; Chiu, M. Differential scanning calorimetry: An invaluable tool for a detailed thermodynamic characterization of macromolecules and their interactions. *J. Pharm. Bioall. Sci.* **2011**, *3*, 39–60. [CrossRef]

55. Chen, Y.; Gao, Z.; Duan, J.Z. Dissolution Testing of Solid Products. In *Developing Solid Oral Dosage Forms*, 2nd ed.; Qiu, Y., Chen, Y., Zhang, G.G.Z., Yu, L., Mantri, R.V., Eds.; Academic Press: New York, NY, USA, 2017; pp. 355–379.

56. Shohin, I.E.; Grebenkin, D.Y.; Malashenko, E.A.; Stanishevskii, Y.M.; Ramenskaya, G.V. A brief review of the FDA dissolution methods database. *Dissolution Technol.* **2016**, *23*, 6–10. [CrossRef]

57. Chuasuwan, B.; Binjesoh, V.; Polli, J.E.; Zhang, H.; Amidon, G.L.; Junginger, H.E.; Midha, K.K.; Shah, V.P.; Stavchansky, S.; Dressman, J.B.; et al. Biowaiver Monographs for immediate release solid oral dosage form: Diclofenac sodium and diclofenac potassium. *J. Pharm. Sci.* **2009**, *98*, 1206–1219. [CrossRef] [PubMed]

58. Nugrahani, I.; Utami, D.; Nugraha, Y.P.; Uekusa, H.; Hasianna, R.; Darusman, A.A. Cocrystal construction between the ethyl ester with parent drug of diclofenac: Structural, stability, and anti-inflammatory study. *Heliyon* **2019**, *5*, e02946. [CrossRef] [PubMed]

59. Wilson, C.R.; Butz, J.K.; Mengel, M.C. Methods for analysis of gastrointestinal toxicants. *Bio. Sci.* **2014**, *2014*, 1–8.

60. Ren, S.; Liu, M.; Hong, C.; Li, G.; Sun, J.; Wang, J.; Xie, Y. The effects of pH, surfactant, ion concentration, coformer, and molecular arrangement on the solubility behavior of myricetin cocrystal. *Acta Pharm. Sin. B* **2019**, *9*, 59–73. [CrossRef] [PubMed]

 *pharmaceutics*

Article

Application of 1-Hydroxy-4,5-Dimethyl-Imidazole 3-Oxide as Coformer in Formation of Pharmaceutical Cocrystals

Aneta Wróblewska [1], Justyna Śniechowska [1], Sławomir Kaźmierski [1], Ewelina Wielgus [1], Grzegorz D. Bujacz [2], Grzegorz Mlostoń [3], Arkadiusz Chworos [1], Justyna Suwara [1] and Marek J. Potrzebowski [1,*]

[1] Centre of Molecular and Macromolecular Studies, Polish Academy of Sciences, Sienkiewicza 112, 90-363 Lodz, Poland; awroblew@cbmm.lodz.pl (A.W.); jsniech@cbmm.lodz.pl (J.Ś.); kaslawek@cbmm.lodz.pl (S.K.); ms@cbmm.lodz.pl (E.W.); achworos@cbmm.lodz.pl (A.C.); jmilczar@cbmm.lodz.pl (J.S.)

[2] Institute of Technical Biochemistry, Lodz University of Technology, Stefanowskiego 4/10, 90-924 Lodz, Poland; grzegorz.bujacz@p.lodz.pl

[3] Department of Organic and Applied Chemistry, University of Lodz, Tamka 12, 91-403 Lodz, Poland; grzegorz.mloston@chemia.uni.lodz.pl

* Correspondence: marekpot@cbmm.lodz.pl; Tel.: +48-42-680-3240

Received: 16 March 2020; Accepted: 10 April 2020; Published: 15 April 2020

Abstract: Two, well defined binary crystals with 1-Hydroxy-4,5-Dimethyl-Imidazole 3-Oxide (HIMO) as coformer and thiobarbituric acid (TBA) as well barbituric acid (BA) as Active Pharmaceutical Ingredients (APIs) were obtained by cocrystallization (from methanol) or mechanochemically by grinding. The progress of cocrystal formation in a ball mill was monitored by means of high-resolution, solid state NMR spectroscopy. The ^{13}C CP/MAS, ^{15}N CP/MAS and ^{1}H Very Fast (VF) MAS NMR procedures were employed to inspect the tautomeric forms of the APIs, structure elucidation of the coformer and the obtained cocrystals. Single crystal X-ray studies allowed us to define the molecular structure and crystal packing for the coformer as well as the TBA/HIMO and BA/HIMO cocrystals. The intermolecular hydrogen bonding, π–π interactions and CH-π contacts responsible for higher order organization of supramolecular structures were determined. Biological studies of HIMO and the obtained cocrystals suggest that these complexes are not cytotoxic and can potentially be considered as therapeutic materials.

Keywords: imidazole *N*-oxides; barbituric acid; thiobarbituric acid; pharmaceutical cocrystals; mechanochemistry; solid state NMR; X-ray Diffraction

---

1. Introduction

Pharmaceutical cocrystals have recently received a great deal of attention because of their therapeutic importance and remaining challenges [1–5]. The commercial success of rationally designed drugs which have been introduced into the market in last decade, such as Steglatro® (a molecular cocrystal of ertugliflozin and L-pyroglutamic acid), Odomzol® (a cocrystal of sonidegib and phosphoric acid), Suglat® (a cocrystal of ipragliflozin and L-proline), and Entresto1® (a cocrystal of valsartan and sacubitril), has prompted "Big Pharma" companies to double their efforts regarding testing the compositions of new solid systems fulfilling the conditions whereby they may be classified as cocrystals.

According to the generally accepted definition, cocrystals are homogenous (single phase) crystalline structures which are made up of two or more components in a definite stoichiometric ratio where the arrangement in the crystal lattice is not based on ionic bonds (as with salts) [6].

The remarkable advantage of pharmaceutical cocrystals is their significant improvement in terms of physicochemical properties without compromising on therapeutic benefit [7]. It is well known that many Active Pharmaceutical Ingredients (APIs) suffer from low solubility and/or poor permeability [8,9]. Notably, up to 90% of new drugs receive a BCS II (Biopharmaceutics Classification System) category rating, meaning that they demonstrate low solubility and high permeability [10]. In the case of cocrystals, these parameters are usually significantly enhanced.

For scientists dealing with solid state matter and the formulation of new crystalline materials, the preparation of cocrystals with desired properties is usually a challenging task. Progress in this field has been made possible due to great achievements in crystal engineering which allow researchers to predict supramolecular interactions, resulting in new solid forms [11]. Generally, therapeutic cocrystals consist of two components: a biologically active organic compound and a complementary molecular coformer. In the pharmaceutical sciences, the correct choice of a coformer for a desired API is essential [12]. The library of available coformers which fulfill structural prerequisites is significant; however, only selected compounds can be considered to belong to the group of "pharmaceutically useful" ones [13–15]. The basic requirement for a suitable coformer is to be pharmaceutically acceptable, i.e., generally regarded as safe (GRAS) substances. Furthermore, coformers should be relatively cheap, with rather low molecular weight, and possess multiple API-binding sites which can be involved in the formation of strong intermolecular interactions [16].

The intention of our work was to introduce a new compound containing the imidazole core to the group of useful pharmaceutical coformers. In the current project, we selected 1-Hydroxy-4,5-Dimethyl-Imidazole 3-Oxide (**1**) 1-Hydroxy-4,5-Dimethyl-Imidazole 3-Oxide (HIMO) [17] (see Figure 1), which has never been tested before in therapeutic applications. This compound, bearing appropriate functional groups, is capable of forming ordered solid structures. Moreover, a strong motivation for its selection was the fact that numerous imidazole derivatives play an important role in organic chemistry, medicinal chemistry as well as the pharmaceutical sciences. Imidazole derivatives with histidine are common in nature, as basic amino acids. Being a part of the structure of protein, histidine is an essential amino acid which is crucial for the catalytic activity of many enzymes. Moreover, imidazole-based drugs exhibit antiviral [18–20], antitumor [21–23], bacteriostatic [24], and antiprotozoal [25,26] activities. They were reported to act as hypotensive agents [27] and selective inhibitors for some kinases [28,29]. Such unique properties prompted us to prepare new cocrystals, which can be also considered as unknown drug–drug binary systems.

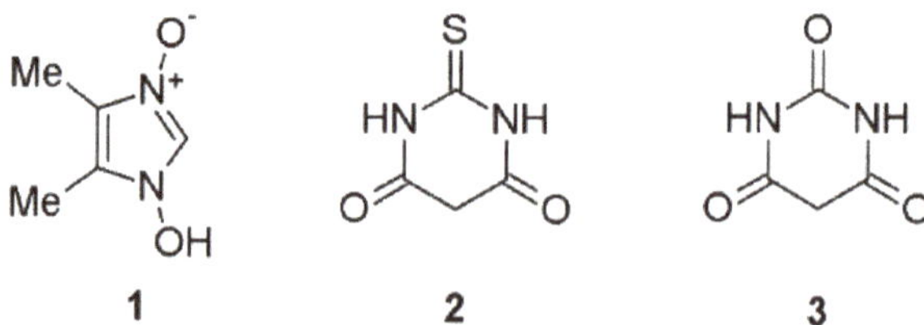

Figure 1. 1-Hydroxy-4,5-Dimethyl-Imidazole 3-Oxide (HIMO) (**1**), thiobarbituric acid (TBA) (**2**) and barbituric acid (BA) (**3**).

The compound 1-hydroxy-4,5-dimethyl-imidazole 3-oxide (**1**) exists in the equilibrium of tautomeric forms [30]. In solution and in gaseous phase, it forms a mixture of the –OH (i.e., 1-hydroxyimidazole) and N→O (i.e., imidazole 3-oxide) tautomers existing in comparable amounts; this fact has been demonstrated by theoretical and experimental studies. The observed equilibria depend not only on the solvent polarity, but also on the composition of substituents attached to the imidazole ring, and their ability to form the inter- or intra- molecular bonds [31–38]. On the other hand, knowledge about the structure of **1** in the solid state is limited, and this makes this compound even more intriguing.

In the present study, thiobarbituric acid (TBA) (**2**) and barbituric acid (BA) (**3**) (Figure 1) were chosen as potential APIs. Both compounds are well known and have been described in numerous crystallographic and physico-chemical reports [38–40].

As highlighted above, the imidazole *N*-oxide **1** has never been tested as a pharmaceutical coformer. Thus, in view of its potential use, we present also biological studies performed for HIMO and its cocrystals with TBA and BA, respectively. Compounds **1**, **2**, **3** and new cocrystals were analyzed using solid-state NMR spectroscopy (SS NMR), single-crystal and powder X-ray Diffraction (XRD) techniques, as well as Differential Scanning Calorimetry (DSC).

2. Materials and Methods

The 1-hydroxy-4,5-dimethyl-imidazole 3-oxide (**1**) was obtained employing the procedure reported in the literature [17]. Compounds **2** and **3** were purchased from Sigma Aldrich (Poznan, Poland) and used without further purification. Reagent grade solvents were purchased from Polish Chemicals Reagents (POCH) company (Gliwice, Poland).

2.1. Preparation of Thiobarbituric Acid/1-Hydroxy-4,5-Dimethyl-Imidazole 3-Oxide (TBA/HIMO) Cocrystal

Method 1

Thiobarbituric acid (**2**) (14.5 mg, 0.10 mmol) and **1** (13.0 mg, 0.10 mmol) were dissolved in boiling MeOH and left for 8 days at room temperature for crystallization. After that, the precipitated TBA/HIMO cocrystals were filtered and used for further investigations.

Method 2

Thiobarbituric acid (**2**) (67.0 mg, 0.47 mmol) and **1** (60.0 mg, 0.47 mmol) were mixed with 0.15 mL of MeOH as a LAG and ground for 1 h in a ball-mill (agate jar—5 mL agate ball—5 mm; 25 Hz). After that, the desired cocrystals were removed from the jar and used for further investigation employing analytical techniques.

2.2. Preparation of Barbituric Acid/1-Hydroxy-4,5-Dimethyl-Imidazole 3-Oxide (BA/HIMO) Cocrystal

Method 1

The barbituric acid (**3**) (13.0 mg, 0.10 mmol) and **1** (13.0 mg, 0.10 mmol) were dissolved in boiling MeOH and left for 5 days at room temperature for crystallization. After that time, the precipitate of BA/HIMO cocrystals was filtered and used for further investigations.

Method 2

The barbituric acid (**3**) (60.0 mg, 0.47 mmol) and **1** (60.0 mg, 0.47 mmol) were mixed with 0.15 mL of MeOH as a LAG and ground for 2 h in a ball-mill (agate jar—5 mL agate ball—5 mm; 25 Hz). After that, the desired cocrystals were removed from jar and investigated employing analytical techniques.

2.3. NMR Spectroscopy

The NMR measurements were carried out using Bruker Avance III 600 and on a Bruker Avance III 400 spectrometers (Bruker GMbH, Karlsruhe, Germany). The Bruker Avance III 600 was equipped with 4 mm ^{1}H/BB (^{31}P–^{15}N) CP/MAS probehead. The resonance frequencies for ^{1}H, ^{13}C, ^{15}N were set at 600.13, 150.90, and 60.82 MHz, respectively. The ^{13}C and ^{15}N CP/MAS spectra were measured with a proton 90° pulse length of 4 µs, a 4–200 s repetition time and SPINAL64 decoupling (83 kHz amplitude). The ^{1}H Very Fast MAS NMR spectra were recorded with a spinning speed of 60 kHz in a 1.3 mm zirconium rotor. Bruker Avance III 400 spectrometer experiments were performed for ^{1}H, ^{15}N and ^{13}C at 400.15, 40.53 and 100.63 MHz resonance frequencies, respectively. The cross polarization magic angle spinning (^{13}C CP MAS) spectra were recorded with a proton 90° pulse length of 4 µs.

In typical CP/MAS experiments, samples were packed into a 4 mm ZrO_2 rotor and spun at a rate of 8–12 kHz. The 1024 transients were acquired. FIDs were accumulated using time domain size of 3.6 K data points. Adamantane (resonances at 38.48 and 29.46 ppm) was used as a secondary ^{13}C chemical-shift reference from external tetramethylsilane (TMS). For ^{15}N, a chemical shift calibration glycine signal of $\delta = 36.6$ ppm was used as a secondary reference.

2.4. Single Crystals X-ray Diffraction Measurements

The HIMO, TBA/HIMO and BA/HIMO crystals were obtained by crystallization from methanol. Single crystal diffraction experiments were carried out using Oxford SuperNova single-crystal diffractometer (Agilent Technologies, Yarnton, UK) with micro-source CuKα radiation ($\lambda + 1.5418$Å) and a Titan detector at room temperature. These conditions were chosen deliberately to show molecular dynamics similar to those present in NMR experiment. The absorption correction was performed based on the crystal shape, orientation and absorption coefficient. Diffraction data collection, cell refinement, data reduction and absorption correction were performed using the CrysAlis PRO software (Oxford Diffraction). Structures were solved by direct method SHELXS and then refined using the full-matrix least-squares method, i.e., SHELX 2015, implemented in the OLEX2 package. The crystals of TBA/HIMO showed pseudo merohedral twinning, which is rare for triclinic systems, possible due to similar *a* and *b* unit cell parameters and α and β cell angles. The refinement was initially performed using one lattice component from diffraction data, and later, the second lattice component was introduced (HKLF 5), which significantly improved the statistical parameters. The hydrogen atoms were set geometrically and refined as riding with the thermal parameter set to 1.2 of the thermal vibration of the parental atom. The dislocated hydrogen atoms were found on the difference Fourier map and refined with 0.5 occupancy for both alternative positions with geometrical restraints and thermal parameters equal to 1.5 of the thermal vibration of the parental atom. These structures were validated by CheckCif (http://checkcif.iucr.org) and deposited in the Cambridge Crystallographic Data Centre (CCDC) under accession numbers 1967291, 19672932 and 1967293 for, TBA/HIMO, BA/HIMO and HIMO and crystals, respectively.

2.5. X-ray Diffraction of Powders

Powder X-ray diffraction experiments were carried out using a Bruker D8 Advance diffractometer (BRUKER AXS, Inc., Madison, WI, USA) equipped with a LYNXEYE position sensitive detector using CuKα radiation ($\lambda = 0.15418$ nm). The data were collected at room temperature in the Bragg–Brentano (θ/θ) geometry, between 3° and 60° (2θ) in a continuous scan using 0.03° steps, under standard laboratory conditions without prior preparation of the measured samples.

2.6. Differential Scanning Calorimetry (DSC) Measurements

DSC measurements were recorded using a DSC 2920, (TA Instruments, New Castle, DE, USA). The heating processes investigated for the TBA/HIMO and BA/HIMO cocrystals were performed in a muffle furnace (SM-2002) manufactured by Czylok (Jastrzebie Zdroj, Poland). The measurements were carried out in temperature range of 0 °C–300 °C. The heating rate was 10 K/min. The sample mass was 1.7040 mg for TBA/HIMO and 1.8200 mg for BA/HIMO.

2.7. Biological Studies of HIMO, TBA/HIMO and BA/HIMO

The cytotoxic impact of the tested compounds on the viability of cells after concentration-dependent treatment was determined by standard MTT [3-(4,5-dimethylthiazol-2-yl)-2,5-diphenyltetrazolium bromide] assay. Experiments were performed with cancer (HeLa–cervical cancer, K562–chronic myelogenous leukemia) and noncancerous (fibroblasts, 293T–derived from human embryonic kidney) human cell lines. A colorimetric assay was used to measure the cell viability through the metabolism of tetrazolium salt into formazan by mitochondrial dehydrogenase in living cells. In brief, the cells were plated into 96-well transparent plates (Nunc) at a density of 7×10^3 cells per well in 200 µL of fresh RPMI or DMEM medium (supplemented with 10% fetal bovine serum and 1% penicillin and streptomycin). After overnight incubation (37 °C, 5% CO_2), the medium was replaced for the control and the wells containing various concentrations (1 μM, 10 μM, 50 μM and 100 μM) of the tested compounds. After a further 48 h of incubation under the same conditions, 25 μL of MTT solution (5 mg/mL) was added to each well. After a subsequent 2 h of incubation, 95 μL of lysis buffer (20% SDS, 50% aqueous DMF, pH 4.5) was added. Afterward, the plates were mixed on a microplate shaker (2 h at room temperature) to dissolve the formazan. Then, the optical density (OD) was measured by a microplate reader at 570 nm with a reference wavelength of 630 nm. Percent cell viability relative to the control was calculated as cell viability (%) = (OD treatment/OD untreated control cells) × 100%. The data represent the mean values from five repeats from three independent experiments.

2.8. Solubility Measurements of BA, TBA, TBA/HIMO and BA/HIMO

Solubility measurements were performed by preparing supersaturated solutions of the tested cocrystals (BA/HIMO, TBA/HIMO) and pure BA and TBA in Milli-Q water (pH 5.7), simulated gastric fluid without pepsin (SGFsp, pH 1.2) and ethanol. SGFsp was prepared by adding 2 g NaCl and 7 mL concentrated HCl to 1000 mL distilled water. The obtained suspensions were stirred at room temperature for 24 h. After filtration through a 0.45 μm PTFE syringe, the filtrates were diluted with water to obtain an appropriate concentration (in the range of UPLC-MS calibration).

The concentration of BA and TBA was determined by an ACQUITY UPLC I-Class chromatography system coupled with SYNAPT G2-Si mass spectrometer equipped with an electrospray source and quadrupole-Time-of-Flight mass analyzer (Waters Corp., Milford, MA, USA). Acquity BEH_{TM} C18 column (100 × 2.1 mm, 1.7 μm), maintained at 45 °C for the chromatographic separation of analyte. A gradient was employed with the mobile phase combining solvent A (1% formic acid in water) and solvent B (1% formic acid in acetonitrile) as follows: 3% B (0–0.5 min), 3–97% B (0.5–1.5 min), 97–97% B (1.5–1.8 min), 97–3% B (1.8–1.9 min) and 3–3% B (1.9–3.5 min). The flow rate was 0.40 mL/min, and the injection volume was 2 μL.

For mass spectrometric detection, the electrospray source operated in negative resolution mode. The optimized source parameters were as follows: capillary voltage 2.8 kV; cone voltage 25 V; desolvation gas flow 900 L/h with a temperature of 450 °C; nebulizer gas pressure 6.5 bar; and source temperature 110 °C. Mass spectra were recorded over an m/z range of 100 to 1200. Mass spectrometer conditions were optimized by direct infusion of the standard solution. The system was controlled using the MassLynx software (Version 4.1), and data processing (peak area integration, construction of the calibration curve) was performed using TargetLynxTM (Waters Corp., Milford, MA, USA).

The initial stock calibration solutions of BA and TBA were prepared at a concentration of approximately 10 mg/mL in water and stored at 4 °C. The stock solutions were diluted (with water) to obtain the target concentrations. The calibration curves were prepared at seven different concentrations of BA and TBA. The calibration curves were linear over a concentration range from 10 ng/mL to 1500 ng/mL with the correlation coefficients of >0.994 and 0.992, respectively, for BA and TBA.

The concentrations determined for BA and TBA (in pure form and in the tested cocrystals) were reported as the average of two replicated experiments for each sample. The injections for each sample were repeated four times.

3. Results

3.1. Solid State NMR Studies of HIMO

One of the aims of the present project was the comparison of two methods of preparing cocrystals, i.e., classic cocrystallization of components with solvent (solvents) use and solvent free ball milling. For the monitoring of the processes employing the ball mill method, knowledge about the structure and properties of HIMO (1) in the solid state was crucial. In our structural studies of 1 in condensed matter, we started with a ^{13}C CP/MAS NMR experiment performed with a spinning rate of 8 kHz at ambient temperature in a 4 mm zirconium rotor. Figure 2a shows spectrum of the sample purified by precipitation. The unexpected feature of the spectrum was its very high resolution of resonances, which is typical for highly crystalline samples with very well organized crystal lattice. Similarly, high quality spectra are usually recorded for molecules which have crystallized in multiple steps. Going further and analyzing the number of resonances both in aromatic (imidazole) and aliphatic (methyl) regions, we concluded that 1 crystallizes during precipitation in space group with Z′ equal 2. The four, clearly cut methyl signals provide unambiguous proof to confirm this hypothesis.

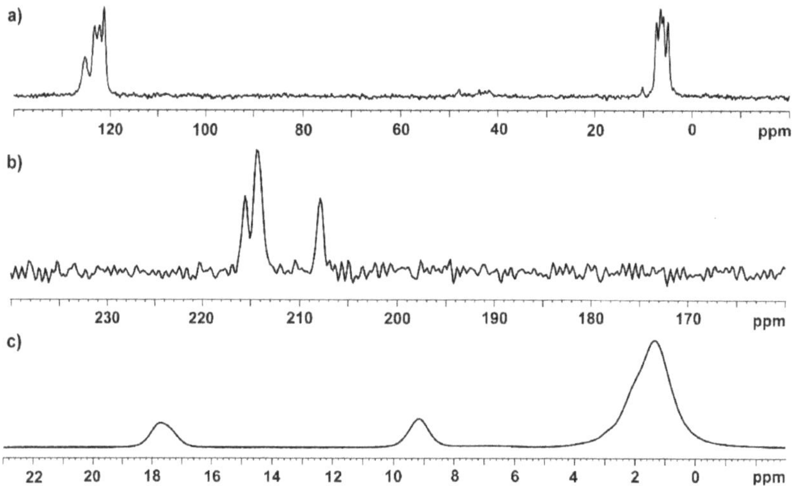

Figure 2. (a) ^{13}C CP/MAS spectrum recorded with a spinning rate of 8 kHz; (b) ^{15}N CP/MAS spectrum recorded with a spinning rate of 8 kHz; (c) ^{1}H VF/MAS NMR spectrum recorded with a spinning rate of 60 kHz.

The ^{15}N CP/MAS spectrum (Figure 2b) confirms the presence of more than one molecule in the asymmetric part of the unit cell. Three clearly cut signals found at δ = 208.0, δ = 214.7 and δ = 216.0 ppm differed in intensity. We assumed that the larger intensity of the middle peak (ca. 215 ppm) resulted from the superposition of two resonance signals representing the A and B molecules.

The ^{1}H NMR spectroscopy is an interesting source of information. For this measurement, we employed a unique technique called Very Fast Magic Angle Spinning (VF MAS). The "very fast" (VF) regime, i.e., at more than 50 kHz, is obtained using commercially available 1.3 mm rotors. This frequency exceeds the strength of the homonuclear ^{1}H dipolar coupling, and is therefore expected to show a new regime for spin dynamics. In many cases, the ^{1}H NMR spectra recorded under VF MAS conditions showed acceptable resolution and the protons positions (chemical shifts) could be correctly assigned.

The ^{1}H VF MAS spectrum (Figure 2c) recorded with spinning rate 60 kHz displays three groups of signals representing protons of methyl residues (δ_{1H} = 1.36 ppm), protons bonded to methine carbon C2 (δ_{1H} = 9.17 ppm) and protons of hydroxyl residue (δ_{1H} = 17.71 ppm). The chemical shifts of *HC*(2) and *OH* are not straightforward. In the first case, the strong deshielding effect was due to the aromatic character of imidazole strengthened by the "zwitterionic" nature of the adjacent N→O unit. In general, the chemical shift at around 9 ppm is characteristic for positions called "acidic". The value of δ_{1H} for *OH* suggests that this residue is involved in a strong hydrogen bonding interaction.

3.2. Formation and Structural Studies of TBA/HIMO and BA/HIMO Cocrystals

3.2.1. SS NMR Studies of TBA/HIMO Cocrystal

The TBA **2** and BA **3**, as well as their derivatives, commonly known as barbiturates, play an important role in pharmaceutical industry because of their hypnotic, sedative, anticonvulsant, antimicrobial, anesthetic, anticancer and antitumor properties [41]. Due to the unique structure of **3** and their ability to form keto–enol tautomers, it is widely used as a valuable building block in organic synthesis and advanced materials including cocrystals [42,43].

In our project, the TBA/HIMO cocrystals were obtained by the cocrystallization of equimolar amounts of TBA and HIMO dissolved in boiling methanol. When the solution was cooled down and kept at ambient temperature for 8 d in a closed vessel, the expected crystals were efficiently formed. After separation of the TBA/HIMO cocrystals, they were dried in air and measured employing ^{13}C CP/MAS, PXRD and DSC techniques (Figure 3). All experiments confirmed the formation of new crystallographic structures. It is worth noting that in the DSC profile (Figure 3c), a strong exothermic peak corresponding to thermal decomposition, typical for high-energetic cocrystals, was observed.

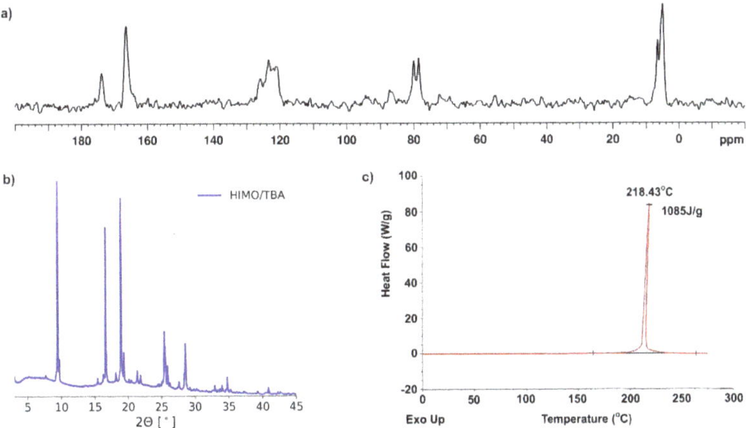

Figure 3. (**a**) ^{13}C CP/MAS spectrum recorded with a spinning rate of 8 kHz for TBA/HIMO cocrystals obtained by cocrystallization from MeOH; (**b**) powder x-ray diffraction diffractogram for TBA/HIMO cocrystals obtained by cocrystallization from MeOH; (**c**) DSC curve of TBA/HIMO cocrystals. For a full analysis of the crystallographic forms of TBA, HIMO and TBA/HIMO based on PXRD diffractograms, see Supporting Information (Figure S1).

It is apparent that in the case of wet methods of cocrystal formation, which are based on the dissolving of starting components, the crystallographic form of applied ingredients is irrelevant. In the case of mechanochemical methods (e.g., ball milling), the structural effects (crystal structure, polymorphism, tautomerism, intermolecular contacts) can affect the process of cocrystal formation. It seems that TBA, for which a rich collection of polymorphs and one hydrated form have been isolated, is an appropriate candidate with which to test the correlation between crystallographic form and the ball milling process. In the crystal lattice, TBA exists in enol form, keto isomer, as well as a mixture of keto/enol forms, each of which can be easily recognized by ^{13}C CP/MAS experiments (see [43]). Depending on the solvent used for the crystallization, it is possible to obtain a suitable form of TBA.

The structure, tautomerism and crystal forms of TBA were investigated by Chierotti et al. [44]. Three different forms crystallized from different solvents were also used as starting materials in the present study. TBA(I) was obtained using MeCN as a solvent for crystallization. This sample exists as tautomeric form A (Scheme 1).

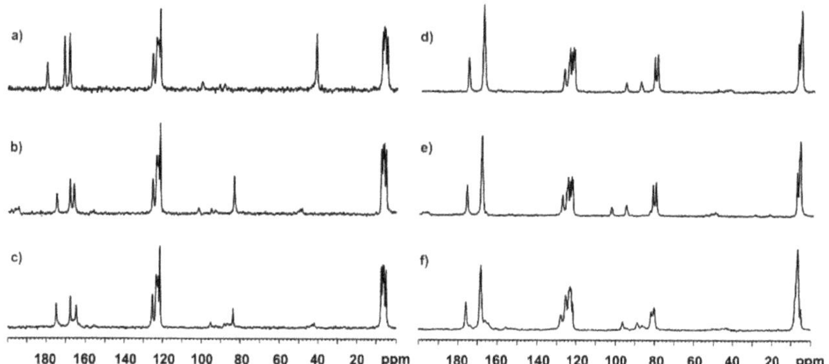

Scheme 1. Numbering system for TBA and tautomers of TBA depending on crystallization from different solvents: tautomeric form A (keto) after the crystallization from MeCN, tautomeric form B (keto/enol) after the crystallization from EtOH or water.

In this form, the C(O)NH moieties on all molecules are either involved in the typical hydrogen-bonding ring motif, with formation of a dimer, or in the formation of hydrogen-bonded chains. TBA (II) was obtained by slow evaporation of a hot ethanolic solution. TBA (II) crystallizes as a tautomer B and forms a type of hydrogen-bonded zigzag chain. Recrystallization of TBA from water leads to the formation of hydrated form III, presented as B, with 1.5 water molecules per formula unit. The acid molecules form a chain through two types of hydrogen-bonding rings involving C(O)NH and C(S)NH moieties.

Figure 4 (left column, a–c) shows the ^{13}C CP/MAS spectra for the physical mixture of two components, i.e., TBA in different forms (I-III) and HIMO, in a 1:1 ratio. It is clear that the intensity of the signals does not reflect the content of the mixture. This is because the relaxation times and efficiency of the cross-polarization for both compounds were significantly different. The TBA/HIMO cocrystals were obtained mechanochemically in the presence of MeOH added as a LAG. Figure 4 (right column, d–f) shows the ^{13}C CP/MAS spectra after one hour of grinding in a Mixer Mill MM 200 equipped with a 5 mL agate jar and 5 mm diameter balls at an oscillation rate of 25 Hz. A comparative analysis of the spectra proved that during the grinding process, in each case, the formation of cocrystals was observed, and the final product was always very similar. The only difference was observed for TBA (III); in that case, a small amount of unreacted HIMO was detected in the crude mixture. Apparently, the 1 h grinding was not sufficient to complete the expected transformation.

Figure 4. The 150.90 MHz ^{13}C CP/MAS spectra recorded with a spinning rate of 11 or 12 kHz. The left column shows a physical mixture of HIMO and TBA in a ratio of 1:1, (**a**) HIMO and TBA crystallized from acetonitrile (tautomer A); (**b**) HIMO and TBA crystallized from EtOH (tautomer B); (**c**) HIMO and TBA crystallized from water (tautomer B). Right column displays the spectra of samples after 1 h of grinding with MeOH as a LAG, (**d**) HIMO and TBA crystallized from MeCN; (**e**) HIMO and TBA crystallized from EtOH; (**f**) HIMO and TBA crystallized from water.

The ^{13}C CP/MAS results suggest that during the mechanosynthesis, each TBA form yields the same product in the presence of HIMO. This hypothesis was further confirmed by ^{15}N CP/MAS experiments. As in the case the of ^{13}C NMR, the left column depicts the spectra of physical mixtures of TBA I-III/HIMO (Figure 5a–c). The TBA I crystallized from MeCN (Figure 5a) is characterized by two sharp peaks at δ = 171.7 and 167.2 ppm, respectively. For TBA II, the chemical shifts were found to be at δ = 169.8 ppm and 160.0 ppm, while for TBA III, the chemical shits were at δ = 174.1 and 157.2 ppm. The distinction in relative intensities of the HIMO versus the TBA ^{15}N signals depended on differences in relaxation times of TBA. The right column (Figure 5d–f) displays the spectra for the obtained cocrystals. An inspection of the ^{15}N CP/MAS data shows apparent resemblances. The number of resonances in the isotropic part suggests that in the asymmetric unit, at least two pairs of TBA/HIMO molecules exist (Z′ = 2).

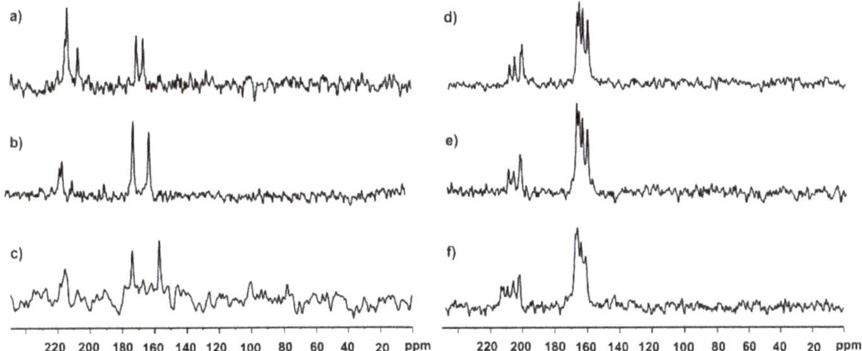

Figure 5. The 60.82 MHz ^{15}N CP/MAS spectra recorded with a spinning rate of 11 or 12 kHz. The left column shows a physical mixture of HIMO and TBA in a ratio of 1:1, (**a**) HIMO and TBA crystallized from MeCN (tautomer A); (**b**) HIMO and TBA crystallized from EtOH (tautomer B); (**c**) HIMO and TBA crystallized from water (tautomer B). The right column displays the spectra of samples after 1 h of grinding with MeOH as a LAG; (**d**) HIMO and TBA crystallized from MeCN; (**e**) HIMO and TBA crystallized from EtOH; (**f**) HIMO and TBA crystallized from water.

The ^{1}H VF/MAS measurements offer an additional source of information for the progress of cocrystal formation. Figure 6 shows the spectra for the mixture of both components (Figure 6a,c,e) and samples after 1 h grinding (Figure 6b,d,f). The ^{1}H chemical shifts for OH and NH protons located in the region 18–10 ppm provide evidence for the very complex hydrogen bonding network in the crystal lattice. The blue vertical line represents the position of OH proton for pure, unreacted HIMO. The small shift of this signal is a probe sensing the course of grinding process. Based on the analysis of spectra for ball milled samples, one can conclude that in experiments with TBA I and TBA II, the transformation was quantitative, while in case of TBA III (crystallized from water), the part of HIMO was not involved in the formation of cocrystals. This observation is consistent with both ^{13}C and ^{15}N analysis.

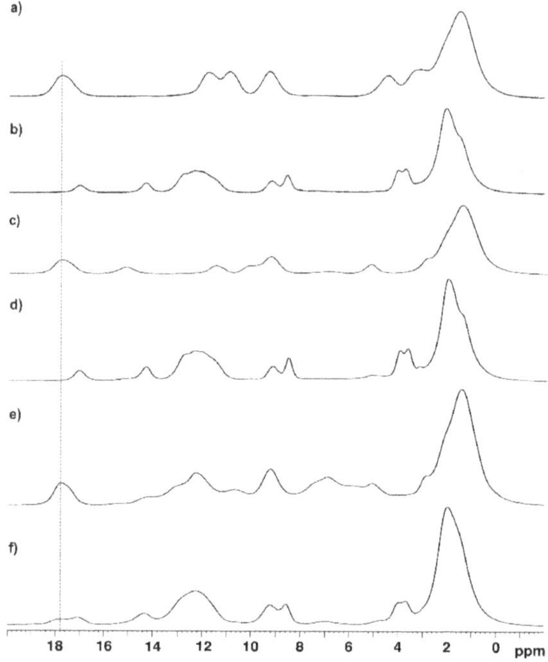

Figure 6. The 600.13 MHz ^{1}H VF/MAS spectra recorded with a spinning rate 60 kHz. Lines a, c and e show a physical mixture of HIMO and TBA in a ratio of 1:1. Lines b, d and f depict the spectra of samples after 1 h of grinding with MeOH as a LAG (**a**) HIMO and TBA crystallized from MeCN, mixture; (**b**) HIMO and TBA crystallized from MeCN, ground; (**c**) HIMO and TBA crystallized from EtOH, mixture; (**d**) HIMO and TBA crystallized from EtOH, ground; (**e**) HIMO and TBA crystallized from water, mixture; (**f**) HIMO and TBA crystallized from water, ground. The vertical blue line shows that the position of the O-H group of HIMO is not involved in interactions with TBA.

3.2.2. SS NMR Structural Studies of BA/HIMO Cocrystal

In the next step, employing wet and solid state methods, we tested the ability of HIMO to form cocrystals with barbituric acid (BA), an analog of TBA. As before, the cocrystallization of BA with HIMO in MeOH led to the formation of the desired BA/HIMO cocrystals. Its structure was confirmed by ^{13}C CP/MAS, PXRD and DSC measurements (Figure 7).

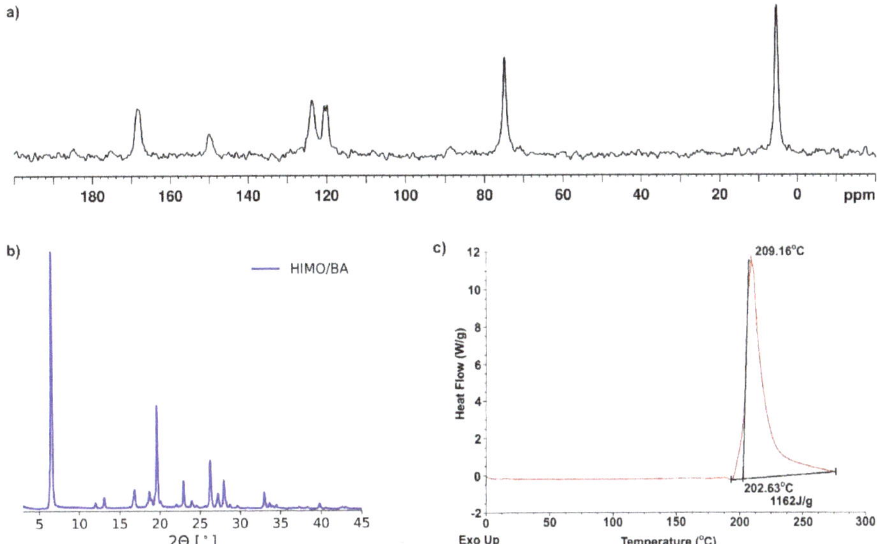

Figure 7. (a) ^{13}C CP/MAS spectrum recorded with a spinning rate of 8 kHz for BA/HIMO cocrystals obtained by cocrystallization from MeOH; (b) powder x-ray diffraction diffractogram for BA/HIMO cocrystal obtained by cocrystallization from MeOH; (c) DSC curve of BA/HIMO cocrystal. For a full analysis of the crystallographic forms of BA, HIMO and BA/HIMO based on PXRD diffractograms, see Supporting Information (Figure S3).

Analyzing the reactivity of BA in mechanochemical procedures, it has to be noted that this compound forms different polymorphs, two anhydrous forms (form I and II, see Scheme 2) and a dihydrate phase. They present a trioxo structure, as confirmed by X-ray diffraction analysis. Recently, Chieriotti and coworkers proved the existence of unstable trihydroxyl and keto-enol tautomers of BA in the crystal lattice [45].

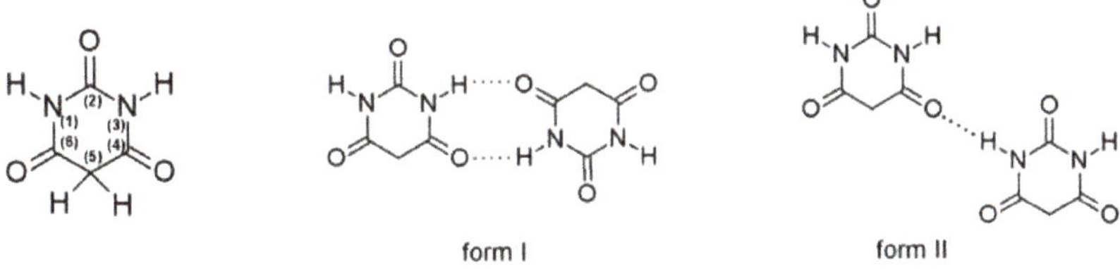

Scheme 2. Numbering system for BA and hydrogen bonding in the form I and II.

Figure 8a shows a mixture of BA and HIMO in a molar ratio of 1:1, crystallized from MeOH (labeled in the literature as Form II). The ^{13}C CP/MAS spectra for BA form II have already been reported [44,45]. Two CH_2 peaks located at 39.1 and at 41.1 ppm represent two independent molecules in the asymmetric unit cell where the CH_2 moiety can be in or out of the plane of the ring in a half chair conformation. The resonances at 151.9, 170.4 and 171.6 (shoulder) ppm were associated with the C2, C6, and C4 carbon atoms of the carbonyl groups, respectively. The assignment of ^{13}C NMR resonances for 1 is presented in Section 3.1. Figure 8c displays the sample prepared mechanochemically (Mixer Mill MM 200 equipped with a 5 mL agate jar and 5 mm diameter ball at an oscillation rate of 25 Hz). The ^{13}C CP/MAS pattern shows changes and provides proof for the formation of a new species. The new signal at 79.5 ppm, which is characteristic for keto/enol-form, is apparent.

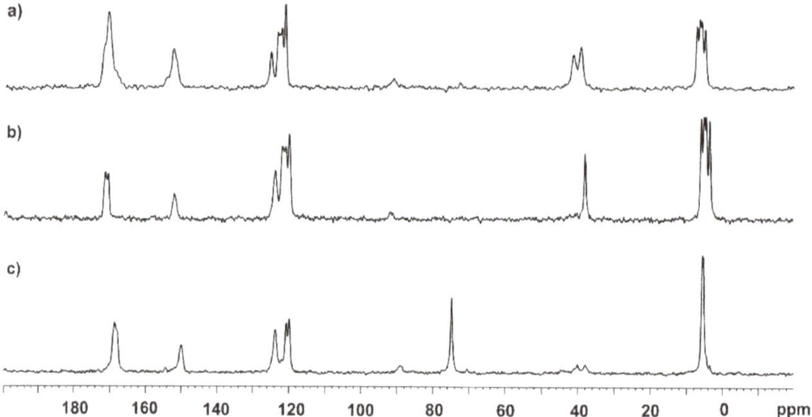

Figure 8. ^{13}C CP/MAS Spectra recorded with a spinning rate of 8 kHz for (**a**) physical mixture of HIMO and BA (II) in a ratio of 1:1; (**b**) physical mixture of HIMO and BA-dihydrate in a ratio of 1:1; (**c**) HIMO and BA in a ratio of 1:1 after 2 h of grinding employing MeOH as LAG.

In extension of this study, we investigated crystals of BA containing two water molecules in the lattice, and the susceptibility of this form to yield cocrystals. Recently, King and coworkers reported on detailed structural studies on, and the unusual thermal phase behavior of, barbituric acid (BA) dihydrate [46]. Our results prove that during grinding at ambient temperature, this form is stable and does not undergo further transformation.

Figure 8b shows the ^{13}C CP/MAS spectrum of the mixture BA·2H$_2$O with HIMO in 1:1 molecular ratio. The peak at $\delta = 40$ ppm is characteristic for the keto form of BA. The cocrystal was prepared by grinding in the Mixer Mill MM 200. In this case, we did not use LAG, assuming that crystalline water could be a substitute. After 1 h ball-milling, the process of BA/HIMO cocrystal formation was complete. The registered ^{13}C NMR spectrum, like those shown in Figure 8c, confirms the structure of product.

The presence of the keto-form of BA further confirmed the validity of the ^{15}N CP/MAS experiment. The spectrum of BA (Figure 9a) with one NMR signal located at $\delta = 150.5$ ppm is consistent with data published by Chierotti et al. [45]. In the case of BA dihydrates, signals at $\delta = 150.5$ and 153.6 ppm were observed (Figure 9b); Figure 9c shows the ^{15}N CP/MAS spectrum of BA:HIMO cocrystal. Analysis of this spectrum suggests that ^{15}N resonance representing the BA component was upfield shifted by ca. 10 ppm ($\delta = 140.2$ ppm) compared to pure BA, while ^{15}N representing HIMO showed a split signal at $\delta = 208.0$ ppm. Figure 9d displays the ^{1}H NMR VF/MAS spectrum of HIMO and BA (ratio 1:1) after 2 h of grinding recorded with a spinning rate 60 kHz. The resolution of this spectrum makes the assignment of proton signals unambiguous. The first striking difference is the lack of the OH signal at $\delta_{1H} = 17.7$ ppm, which, in pure sample 1, represents strong O-H···O hydrogen bonding and the appearance of new signals at $\delta = 15.7$, 14.8 and $\delta = 9.6$ ppm. The collected NMR data are a valuable source of information about the organization of new materials (HIMO and BA in a ratio of 1:1 after 2 h of grinding) and about the tautomeric form of BA.

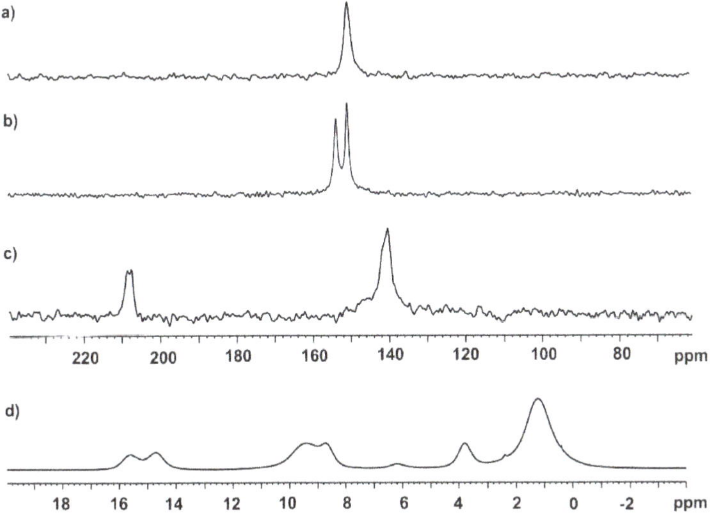

Figure 9. ^{15}N CP/MAS Spectra recorded with a spinning rate of 8 kHz for (**a**) barbituric acid; (**b**) barbituric acid dihydrate; (**c**) HIMO and BA in a ratio of 1:1 after 2 h of grinding; (**d**) ^{1}H VF/MAS spectrum recorded with a spinning rate of 60 kHz for HIMO and BA in a ratio of 1:1 after 2 h of grinding.

3.3. X-ray Structure of HIMO and TBA/HIMO, BA/HIMO Cocrystals

Pure HIMO crystallizes in the triclinic system in the centrosymmetric P-1 space group. The asymmetric unit contains two molecules of HIMO (Z′ = 2). The crystal structure is displayed in Figure 10. Table 1 shows the experimental details and structural information.

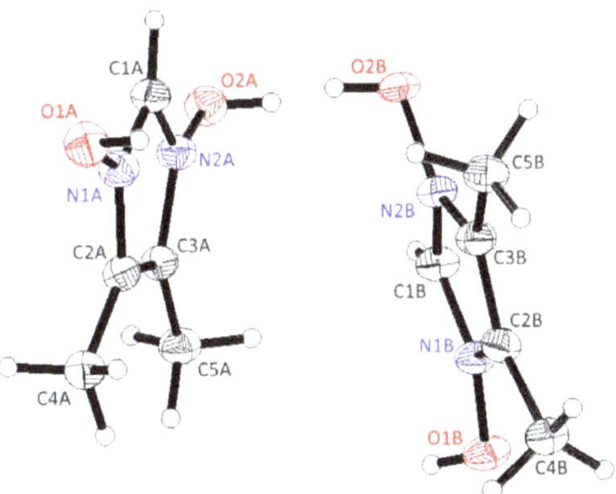

Figure 10. Ortep drawing of the independent unit with a numbering system for HIMO; the position of the hydrogen atom is averaged between both oxygen acceptors at two locations with half occupancy.

The supramolecular array depicted in Figure 11 shows two structural features which are worthy of mention. Firstly, there are two strong hydrogen bonds with O···O distances equal 2.437 Å and 2.446 Å between molecules, creating asymmetric units and molecules related by a symmetry operation. These contacts are responsible for the formation of the chain structure. In both cases, the position of the hydrogen atom is averaged between both oxygen acceptors at two locations with half occupancy.

The second feature seen in Figure 11 is a π–π interaction which is an additional factor stabilizing the crystal structure.

Table 1. Crystal parameters and refinement statistics.

Crystal	HIMO	TBA/HIMO	BA/HIMO
Empirical formula	$C_5H_8N_2O_2$	$C_5H_8N_2O_2 + C_4H_3N_2O_2S$	$C_5H_8N_2O_2 + C_4H_3N_2O_3$
Formula weight	128.14	272.28	256.23
Temperature	293 K	293 K	173 K
Crystal system	Triclinic	Triclinic	Triclinic
Space group	P-1	P-1	P-1
a (Å)	7.5947(4)	10.7777(2)	5.4971(4)
b (Å)	8.5233(5)	11.0568(3)	7.6116(5)
c (Å)	9.7643(4)	11.9869(2)	13.9014(11)
α (°)	85.772(4)	97.968(2)	93.952(6)
β (°)	78.587(4)	97.799(2)	98.627(7)
γ (°)	83.727(5)	118.738(3)	101.157(6)
Volume (Å)^3	614.97(6)	1206.26(5)	561.32(8)
Z	4	4	2
Z′	2	2	1
Density (g·cm^{-3})	1.384	1.499	1.510
Θ Range (°)	4.63 to 75.46	3.79 to 75.25	3.23 to 74.96
Index ranges	$-9 \leq h \leq 9, -10 \leq k \leq 10,$ $-12 \leq l \leq 12$	$-13 \leq h \leq 13, -13 \leq k \leq 13,$ $-13 \leq l \leq 13$	$-6 \leq h \leq 6, -9 \leq k \leq 9,$ $-17 \leq l \leq 17$
Nref	2489	4726	2268
R (reflections)	0.0494 (2300)	0.0535 (4485)	0.0899 (1825)
wR$_2$ (reflections)	0.1570 (2489)	0.1745 (4726)	0.2963 (2268)

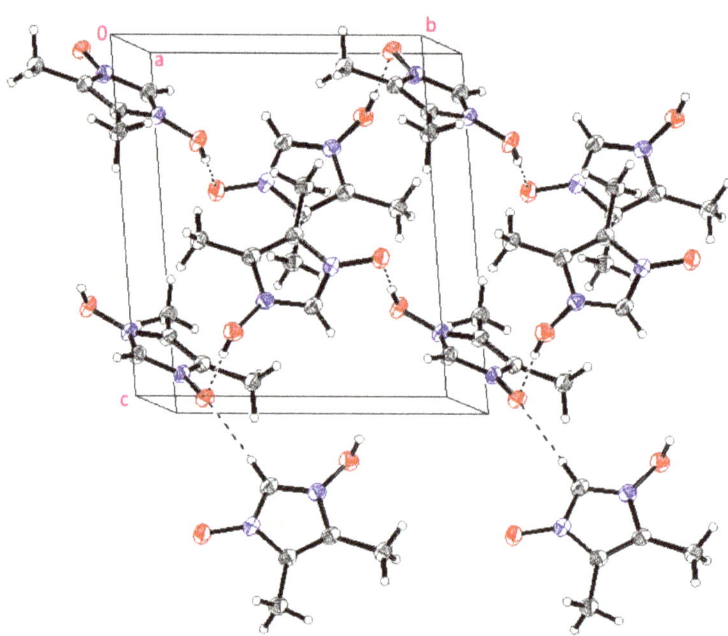

Figure 11. Molecular packing of HIMO.

The crystal structure and molecular packing of TBA/HIMO is shown in Figure 12. The experimental details and structural information are collected in Table 1. The TBA/HIMO crystallizes in the triclinic system in the centrosymmetric P-1 space group. The independent unit contains two molecules of TBA and two molecules of HIMO. Both components in the crystal lattice are connected by hydrogen bonds forming a unique supramolecular structure. The leading motif is created by the planary located TBA molecules forming ribbons (flat chains), rotated such that the thiocarbonyl groups point into the opposite directions. These chains are bonded by strong hydrogen bridges between $C_C(2)=O_C(1)\cdots H_D(2)-N_D(2)$ and $C_C(4)=O_C(2)\cdots H_D(1)-N_D(1)$, and further by $C_D(4)=O_D(2)\cdots H_C(1)-N_C(1)$ and $C_D(2)=O_D(1)\cdots H_C(2)-N_C(2)$. TBA chains are connected by the HIMO molecules to create a plane structure. The methyl groups of HIMO from two layers are pointing into the plane interface. The interaction in other interfaces between planes is created by π stacking interactions.

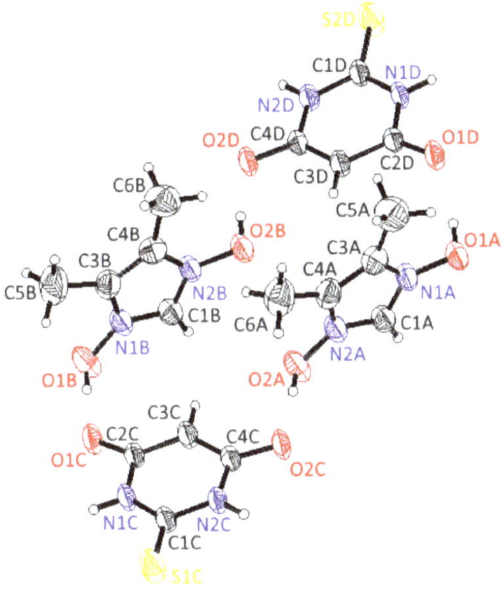

Figure 12. Ortep drawing of the independent unit in TBA/HIMO crystal.

Figure 13 shows the supramolecular pattern for TBA/HIMO in the crystal lattice. Clearly visible characteristic motifs are sheets formed by TBA molecules. These planes are created from the ribbons described above. The distances between $O_C(1)\cdots N_D(2)$ and $O_C(2)\cdots N_D(1)$ were found to be 2.854 Å and 2.839 Å, respectively. The analogous $O_D(2)\cdots N_C(1)$ and $O_D(1)\cdots N_C(2)$ lengths are equal to 2.850 Å and 2.886 Å. The oxygen atoms O1 and O2 of molecules A and B are involved in bifurcated hydrogen bonding with neighboring HIMO molecules. The strength of these bonds is defined by short $O\cdots O$ contacts equal 2.432 Å, 2.598 Å and 2.503 Å, 2.595 Å, respectively. The imidazole *N*-oxide molecules act as a link between adjoining TBA sheets during the formation of hydrogen bridges. As we postulated, at this stage, one of the hydrogens of the methylene group is transferred to the oxygen, then to HIMO oxygen forming intermolecular bonding. The second oxygen acts as an acceptor by interacting with OH residue in HIMO. In such an arrangement, carbonyl C=O groups, which are involved in strong hydrogen bonding, disturb the tautomeric nature of TBA. The TBA loses its pure keto tautomeric form and exists in keto-enol $C-O\cdots H-O-N$ form. Such a complex transfer process makes it difficult to refine the position of the hydrogen in the bridges but position of peak on electron density map shows that is predominantly attached to oxygen in HIMO.

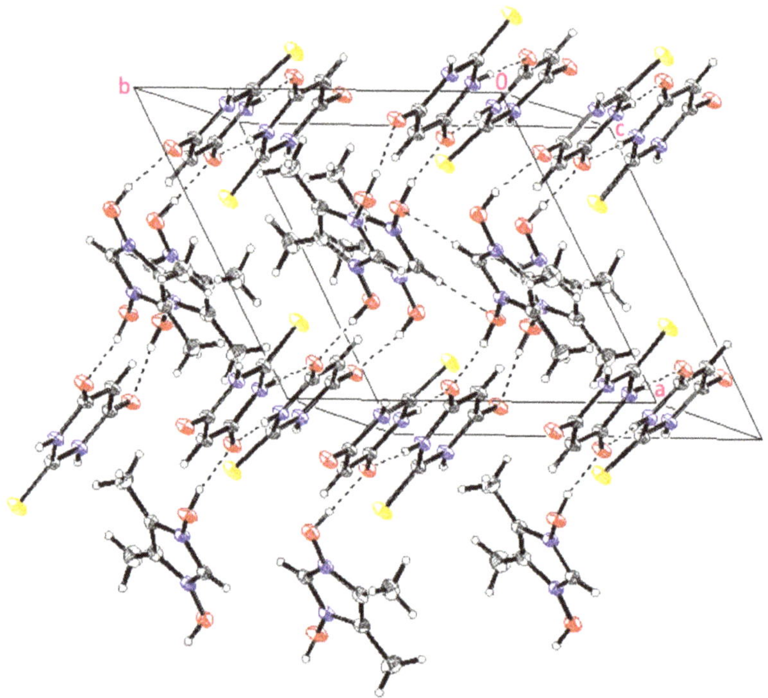

Figure 13. Supramolecular structure of TBA/HIMO cocrystal.

Finally, the supramolecular structure of TBA/HIMO is supplemented by CH⋯π interactions between the methyl groups of HIMO and the ring of TBA. The average distance between the planes of the HIMO and TBA sheets, measured as the distance between the CH_3 atom and the center of the TBA ring, is ca. 4 Å. The imidazole rings are in a stacked arrangement, and are twisted with respect to the TBA plane. It is worth noting that the C=S moiety, in this crystal form, interacts with the π-electrons of imidazole. The distance between the sulfur and the center of the imidazole ring is 3.5 Å.

Figure 14 shows the crystal structure and molecular packing of the BA/HIMO cocrystals. The experimental details and structural information are collected in Table 1. The BA/HIMO sample crystallizes in the triclinic system in the centrosymmetric P-1 space group. The asymmetric unit contains one molecule of BA and one of HIMO (Z = 2). The supramolecular structure is depicted in Figure 15. The motifs seen in the TBA/HIMO structure are also observed for BA/HIMO. The first similar pattern is the ribbon structure (flat chains) formed by barbituric acid. The hydrogen bonding network is similar to the one observed before. The layers are formed by BA ribbons located perpendicularly to the plane and stabilized by π stacking interactions. The HIMO molecules bridge the layers of BA such that protonated oxygens create hydrogen bonds with keto groups of BA. The lengths of N-H⋯O=C bonds were found to be 2.869 Å and 2.912 Å, respectively. The C(1)=O(1) and C(4)=O(3) groups are involved in the formation of bifurcated hydrogen bonds. The counter partner in these bridges is HIMO, which acts as a staple connecting the BA sheets located in parallel planes. The O⋯O contacts are equal to 2.474 Å and 2.593 Å. The keto-enol form is created according to the mechanism described above for TBA, and is also effective for BA/HIMO cocrystals. The evidence confirming the hypothesis that the tautomeric form is forced by hydrogen bonds is found within the analysis of the C-O bond lengths for BA. For C(2)–O(2), this length is 1.222 Å, while for C(1)=O(1) and C(4)=O(3), it is 1.283 Å and 1.276 Å, respectively. It is interesting to note that for pure anhydrous BA, existing in keto form, these bond lengths are 1.229 Å, 1.222 Å, 1.189 Å for C(2)=O(2), C(1)=O(1) and C(4)=O(3), respectively.

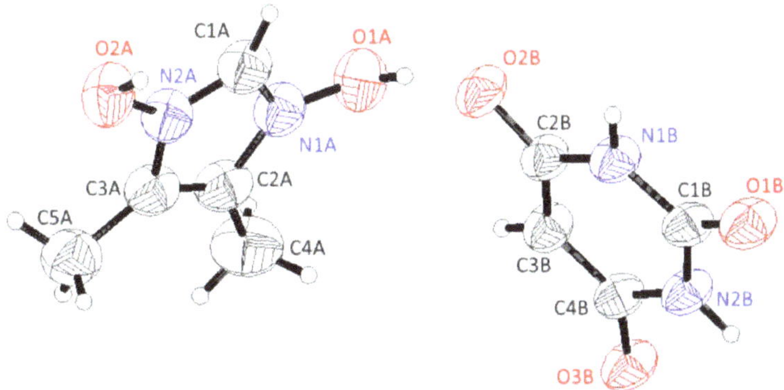

Figure 14. Ortep drawing of the independent unit in BA/HIMO cocrystal.

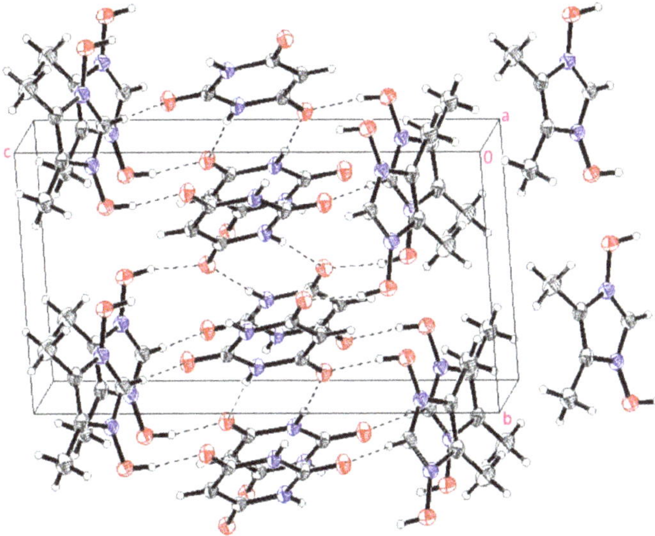

Figure 15. Supramolecular structure of BA/HIMO cocrystal.

The sheets formed by BA are located in parallel planes with a distance of ca. 4 Å. The HIMO clips next to the C(1)=O(1) and C(4)=O(3) BA positions, responsible for creating of structural network, are oriented in opposite directions. Such an orientation allows the BA sheets to join in a "zipper" type mechanism.

Concluding this part, the presented X-ray results are consistent with a publication by Braga and coworkers, who, while studying the gas–solid reactions between the different polymorphic modifications of barbituric acid and amines, reported similar structural motifs with complex hydrogen bonding networks [47].

3.4. Cell Cytotoxicity and Solubility of Tested Compound (HIMO and Its Cocrystals)

As highlighted in the Introduction, 1-hydroxy-4,5-dimethyl-imidazole 3-oxide (**1**) has never been considered as a coformer in the formation of pharmaceutical cocrystals. Among the main problems were a lack of biological tests and limited knowledge about its cytotoxicity.

In view of its potential medical application, cell viability should to be tested to exclude potential toxicity. The cytotoxicity of the studied compounds was measured by the MTT assay method against HeLa (cervical cancer), K562 (chronic myelogenous leukemia) and noncancerous, 293T (derived from human embryonic kidney). Figure 16 shows the cellular viability after concentration-dependent treatment; the results indicate no significant toxicity in the case of incubation with the tested compounds in the 1–100 µM concentration range. The viability of cells was at a level of 80–100 percent, and only in the case of the highest concentration of the first cocrystal (BA/HIMO) did the survival rate decrease to 70% with HeLa cells. The results for noncancerous 293T were similar to those for HeLa cells. The obtained data indicate that all the tested compounds are suitable candidates for use in live cancer and noncancerous cells at concentrations of 1 µM to 100 µM.

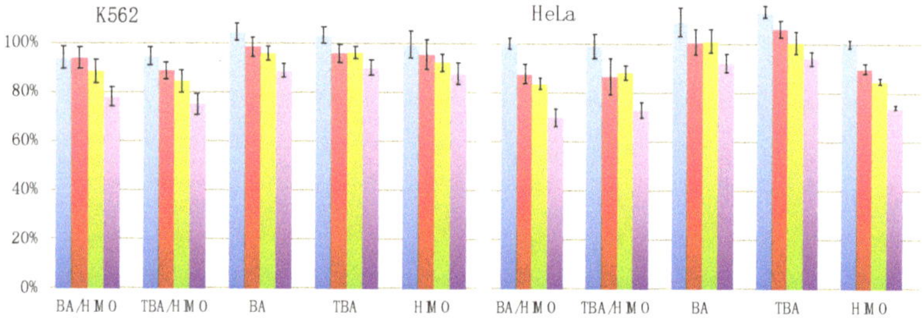

Figure 16. The viability (%) of K562 and HeLa cells after 48 h incubation with the tested compounds in a concentration range from 1 µM (blue bar) to 10 µM, 50 µM and 100 µM (red, green violet bars) respectively. The results represent the mean ± standard error.

Figure 17 shows the solubility of pure TBA and BA samples versus the solubility of the studied cocrystals. The measurements were carried out in three media: water, SGFsp and EtOH. It can be concluded that in water (pH 5.7), the solubility of BA/HIMO cocrystal is slightly better compared to that of BA. In the case of TBA and TBA/HIMO, the relationship is reversed. The solubility of the cocrystals is reduced by over 30%. A cocrystal analysis in simulated gastric fluid without pepsin (SGFsp, pH 1.2) revealed a decrease in solubility for both binary systems compared to TBA and BA. For TBA/HIMO, solubility is reduced by ca. 17%, while for BA/HIMO, it was reduced by about 33% compared to the same cocrystals in water. In EtOH, the solubility of BA is higher compared to that of the cocrystal, while that of TBA and TBA/HIMO is comparable. Comparing the APIs and cocrystals, an increase in solubility was not observed. Usually, increased solubility is the first indication to begin the preparation of cocrystals, because this parameter has an influence on drug pharmacodynamics and the efficiency of treatment. In the case of TBA/HIMO and BA/HIMO, only a minute increase of solubility was noticed for BA/HIMO cocrystals in water. However, while solubility is an important aspect, this is not an arbitrary factor justifying the use of cocrystals in drug delivery. Other functions, for instance, the protection of sensitive APIs from the environmental effects or controlling the delivery of drugs to specific points in the body are important as well. Also relevant from a drug delivery point of view is providing more than one active pharmaceutical ingredient. This seems to be an attractive feature in the case of imidazole *N*-oxide 1 (HIMO).

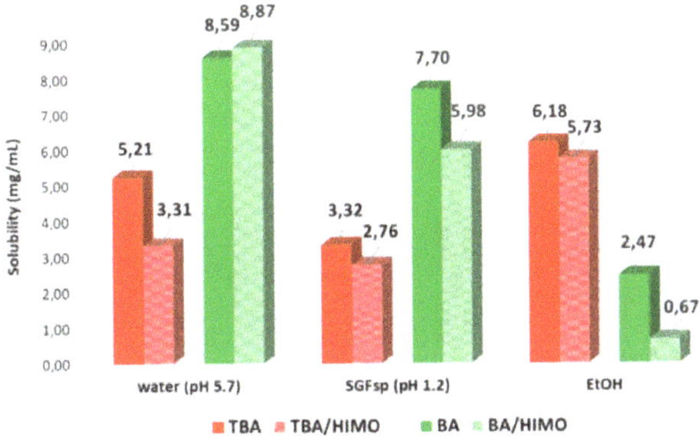

Figure 17. Results of solubility measurements of TBA and BA and their cocrystals with HIMO.

4. Conclusions

This study demonstrates the first application of HIMO as a useful coformer for the preparation of pharmaceutical cocrystals. As API references, thiobarbituric (TBA) and barbituric (BA) acids were employed. Two different methods for the cocrystals preparation were applied: 1) cocrystalization from MeOH used as a solvent, and 2) the grinding of solid components in a ball mill. In both cases, the desired cocrystals (TBA/HIMO and BA/HIMO) were obtained. An advantage of the ball-mill method is that in a classic, "wet" method, a substantial volume of organic solvents has to be used, and the procedure is relatively long (i.e., 8 days). Another important advantage of the ball milling method is its remarkable reduction of the reaction time required to complete the preparation of the cocrystals. Moreover, the ball milling procedure does not require toxic solvents, which is an important feature in view of economic and ecological concerns. The application of diverse analytical methods such as solid-state NMR spectroscopy, X-ray crystallography of single crystals and powders, and IR spectroscopy allowed us to perform an extended structural analysis of the obtained cocrystals. Solid state NMR techniques were used to monitor the progress of cocrystal formation. In addition, it allowed us to evaluate whether the polymorphic and tautomeric forms of the used components influenced the kinetics of the cocrystal formation and the structures of the final products. It was demonstrated that in each case, identical cocrystals formed, irrespective of the applied method (solvent versus ball milling). However, in the latter procedure, TBA hydrate requires a longer processing time.

The biocompatibility for all investigated compounds was also investigated. Experiments performed with cancer (HeLa, K562) and noncancerous (293T) human cell lines suggest that none of tested compounds is toxic. The obtained results strongly suggest that both TBA/HIMO and BA/HIMO cocrystals can be considered potential candidates for pharmaceutical application at a wide range of concentrations, i.e., varying from 1 μM to at least 50 μM. The solubility of the obtained cocrystals was tested in three media. Although only a small increase of solubility was observed for BA/HIMO in water, it seems obvious that the other characteristics of the studied cocrystals (protection of APIs and providing more than one API) may be useful for the development of new drug delivery systems based on 1-hydroxy imidazole 3-oxide derivatives.

Supplementary Materials: The following are available online at http://www.mdpi.com/1999-4923/12/4/359/s1, Figure S1. Calculated powder x-ray diffraction diffractograms (from top to bottom) TBA crystallized from acetonitryl (tautomer A)1; TBA crystallized from ethanol (tautomer B)1; TBA crystallized from water (tautomer B)1; HIMO; TBA/HIMO. Figure S2. 13C CP/MAS spectra recorded with a spinning rate of 8 kHz for TBA/HIMO cocrystals obtained by (a) grinding; (b) cocrystallization from MeOH. Figure S3. Calculated powder x-ray diffraction diffractograms (from top to bottom) for; BA (form II)2, BA dihydrate3; HIMO and BA/HIMO. Figure S4. 13C CP/MAS spectra recorded with a spinning rate of 8 kHz for BA/HIMO cocrystals obtained by (a) grinding; (b) cocrystallization from MeOH.

Author Contributions: Data curation, J.Ś., S.K., E.W. and J.S.; Formal analysis, G.M.; Investigation, A.C.; Methodology, G.D.B.; Project administration, A.W.; Writing—original draft, M.J.P. All authors have read and agreed to the published version of the manuscript.

Funding: This project was financially supported by the National Science Center, Poland (Grants: No. UMO-2017/25/B/ST4/02684, M.P., and UMO-2016/23/G/ST5/04115/l, G.M.).

Conflicts of Interest: The authors declare no conflict of interest.

References

1. Shan, N.; Zaworotko, M.J. The role of cocrystals in pharmaceutical science. *Drug Discov. Today* **2008**, *13*, 440–446. [CrossRef]

2. Kumar, S.; Nanda, A. Approaches to design of pharmaceutical cocrystals: A Review. *Mol. Cryst. Liq. Cryst.* **2018**, *667*, 54–77. [CrossRef]

3. Shaikh, R.; Singh, R.; Walker, G.M.; Croker, D.M. Pharmaceutical cocrystal drug products: An outlook on product development. *Trends Pharmacol. Sci.* **2018**, *39*, 1033–1048. [CrossRef]

4. Kulla, H.; Michalchuk, A.A.L.; Emmerling, F. Manipulating the dynamics of mechanochemical ternary cocrystal formation. *Chem. Commun.* **2019**, *55*, 9793–9796. [CrossRef]

5. Sinha, A.S.; Maguire, A.R.; Lawrence, S.E. Cocrystallization of nutraceuticals. *Cryst. Growth Des.* **2015**, *15*, 984–1009. [CrossRef]

6. Kavanagh, O.N.; Croker, D.M.; Walker, G.M.; Zaworotko, M.J. Pharmaceutical cocrystals: From serendipity to design to application. *Drug Discov. Today* **2019**, *24*, 796–804. [CrossRef]

7. Wang, N.; Xie, C.; Lu, H.; Guo, N.; Lou, Y.; Su, W.; Hao, H. Cocrystal and its Application in the Field of Active Pharmaceutical Ingredients and Food Ingredients. *Curr. Pharm. Des.* **2018**, *24*, 2339–2348. [CrossRef]

8. Good, D.J.; Rodríguez-Hornedo, N. Solubility advantage of pharmaceutical cocrystals. *Cryst. Growth Des.* **2009**, *9*, 2252–2264. [CrossRef]

9. Wang, Y.; Ren, F.; Cao, D. A dynamic and electrostatic potential prediction of the prototropic tautomerism between imidazole 3-oxide and 1-hydroxyimidazole in external electric field. *J. Mol. Model.* **2019**, *25*, 330. [CrossRef]

10. Gadade, D.D.; Pekamwar, S.S. Pharmaceutical cocrystals: Regulatory and strategic aspects, design and development. *Adv. Pharm. Bull.* **2016**, *6*, 479–494. [CrossRef]

11. Sathisaran, I.; Dalvi, S.V. Engineering cocrystals of poorly water-soluble drugs to enhance dissolution in aqueous medium. *Pharmaceutics* **2018**, *10*, 108. [CrossRef]

12. Ross, S.A.; Lamprou, D.A.; Douroumis, D. Engineering and manufacturing of pharmaceutical cocrystals: A review of solvent-free manufacturing technologies. *Chem. Commun.* **2016**, *52*, 8772–8786. [CrossRef]

13. Karagianni, A.; Malamatari, M.; Kachrimanis, K. Pharmaceutical cocrystals: New solid phase modification approaches for the formulation of APIs. *Pharmaceutics* **2018**, *10*, 18. [CrossRef]

14. Paul, M.; Desiraju, G.R. From a binary to a quaternary cocrystal: An unusual supramolecular synthon. *Angew. Chem. Int. Ed.* **2019**, *58*, 12027–12031. [CrossRef]

15. Fukte, S.R.; Wagh, M.P.; Rawat, S. Coformer selection: An important tool in cocrystal formation. *Int. J. Pharm. Pharm. Sci.* **2014**, *6*, 9–14.

16. Karimi-Jafari, M.; Padrela, L.; Walker, G.M.; Croker, D.M. Creating cocrystals: A review of pharmaceutical cocrystal preparation routes and applications. *Cryst. Growth Des.* **2018**, *18*, 6370–6387. [CrossRef]

17. Bartz, S.; Blumenröder, B.; Kern, A.; Fleckenstein, J.; Frohnapfel, S.; Schatz, J.; Wagner, A. Hydroxy-1*H*-imidazole-3-oxides—Synthesis, kinetic acidity, and application in catalysis and supramolecular anion recognition. *Z. Naturforsch.* **2009**, *64b*, 629–638. [CrossRef]

18. Zheng, Z.; Wu, T.; Zheng, R.; Wu, Y.; Zhou, X. Study on the synthesis of quaternary ammonium salts using imidazolium ionic liquid as catalyst. *Catal. Commun.* **2007**, *8*, 39–42. [CrossRef]

19. Zhang, L.; Peng, X.-M.; Damu, G.L.V.; Geng, R.-X.; Zhou, C.-H. Comprehensive Review in current developments of imidazole-based medicinal chemistry. *Med. Res. Rev.* **2013**, *34*, 340–437. [CrossRef]

20. Vekariya, R.L. A review of ionic liquids: Applications towards catalytic organic transformations. *J. Mol. Liq.* **2017**, *227*, 44–60. [CrossRef]

21. Zheng, D.; Dong, L.; Huang, W.; Wu, X.; Nie, N. A review of imidazolium ionic liquids research and development towards working pair of absorption cycle. *Renew. Sustain. Energy Rev.* **2014**, *37*, 47–68. [CrossRef]

22. Pieczonka, A.M.; Strzelczyk, A.; Sadowska, B.; Mlostoń, G.; Stączek, P. Synthesis and evaluation of antimicrobial activity of hydrazones derived from 3-oxido-1*H*-imidazole-4-carbohydrazides. *Eur. J. Med. Chem.* **2013**, *64*, 389–395. [CrossRef]

23. Ghandi, K.A. A review of ionic liquids, their limits and applications. *Green Sustain. Chem.* **2014**, *4*, 44–53. [CrossRef]

24. Irge, D.D. Ionic liquids: A review on greener chemistry applications, quality ionic liquid synthesis and economical viability in a chemical processes. *Am. J. Phys. Chem.* **2016**, *5*, 74–79. [CrossRef]

25. Welton, T. Ionic liquids in catalysis. *Coord. Chem. Rev.* **2004**, *248*, 2459–2477. [CrossRef]

26. Olivier-Bourbigou, H.; Magna, L.; Morvan, D. Ionic liquids and catalysis: Recent progress from knowledge to applications. *Appl. Catal. Gen.* **2010**, *373*, 1–56. [CrossRef]

27. Zhao, D.; Wu, M.; Kou, Y.; Min, E. Ionic liquids: Applications in catalysis. *Catal. Today* **2002**, *74*, 157–189. [CrossRef]

28. Hopkinson, M.N.; Richter, C.; Schedler, M.; Glorius, F. An overview of N-heterocyclic carbenes. *Nature* **2014**, *510*, 485–496. [CrossRef]

29. Fèvre, M.; Pinaud, J.; Gnanou, Y.; Vignolle, J.; Taton, D. N-Heterocyclic carbenes (NHCs) as organocatalysts and structural components in metal-free polymer synthesis. *Chem. Soc. Rev.* **2013**, *42*, 2142–2172. [CrossRef]

30. Nikitina, P.A.; Koldaeva, T.Y.; Mityanov, V.S.; Miroshnikov, V.S.; Basanova, E.I.; Perevalov, V.P. Prototropic tautomerism and some features of the IR spectra of 2-(3-Chromenyl)-1-hydroxyimidazoles. *Aust. J. Chem.* **2019**, *72*, 699–708.

31. Visbal, R.; Gimeno, M.C. N-heterocyclic carbene metal complexes: Photoluminescence and applications. *Chem. Soc. Rev.* **2014**, *43*, 3551–3574. [CrossRef] [PubMed]

32. Luca, L.D. Bacterial symbionts: Prospects for the sustainable production of invertebrate-derived pharmaceuticals. *Curr. Med. Chem.* **2006**, *13*, 1–50. [PubMed]

33. Verma, A.; Joshi, S.; Singh, D. Imidazole: Having versatile biological activities. *J. Chem.* **2013**, *2013*, 1–12. [CrossRef]

34. Menon, K.; Mousa, A.; de Courten, B. Effects of supplementation with carnosine and other histidine-containing dipeptides on chronic disease risk factors and outcomes: Protocol for a systematic review of randomized controlled trials. *BMJ Open* **2018**, *8*, e020623. [CrossRef] [PubMed]

35. Xiao, H.; Peters, F.B.; Yang, P.-Y.; Reed, S.; Chittuluru, J.R.; Schultz, P.G. genetic incorporation of histidine derivatives using an engineered pyrrolysyl-tRNA synthetase. *ACS Chem. Biol.* **2014**, *9*, 1092–1096. [CrossRef]

36. Nikitina, G.V.; Pevzner, M.S. Imidazole and benzimidazole N-oxides (a review). *Chem. Heterocycl. Compd.* **1993**, *29*, 127–151. [CrossRef]

37. Mlostoń, G.; Jasiński, M.; Wróblewska, A.; Heimgartner, H. Recent progress in the chemistry of 2-unsubstituted 1*H*-Imidazole 3-Oxides. *Curr. Org. Chem.* **2016**, *20*, 1359–1369. [CrossRef]

38. Mlostoń, G.; Jasiński, M. First synthesis of the N(1)-bulky substituted imidazole 3-oxides and their complexation with hexafluoroacetone hydrate. *ARKIVOC* **2011**, *6*, 162–175.

39. Chierotti, M.R.; Ferrero, L.; Garino, N.; Gobetto, R.; Pellegrino, L.; Braga, D.; Grepioni, F.; Maini, L. The richest collection of tautomeric polymorphs: The Case of 2-thiobarbituric acid. *Chem. Eur. J.* **2010**, *16*, 4347–4358. [CrossRef]

40. Lewis, T.C.; Tocher, D.A.; Price, S.L. An experimental and theoretical search for polymorphs of barbituric acid: The challenges of even limited conformational flexibility. *Cryst. Growth Des.* **2004**, *4*, 979–987. [CrossRef]

41. Zuccarello, F.; Buemi, G.; Gandolfo, C.; Contino, A. Barbituric and thiobarbituric acids: A conformational and spectroscopic study. *Spectrochim. Acta Part A* **2003**, *59*, 139–151. [CrossRef]

42. Golovnev, N.N.; Molokeev, M.S.; Lesnikov, M.K.; Atuchin, V.V. Two salts and the salt cocrystal of ciprofloxacin with thiobarbituric and barbituric acids: The structure and properties. *J. Phys. Org. Chem.* **2018**, *31*, e3773. [CrossRef]

43. Ziarani, G.M.; Alealia, F.; Lashgari, N. Recent applications of barbituric acid in multicomponent reactions. *RSC Adv.* **2016**, *6*, 50895–50922. [CrossRef]

44. Chierotti, M.R.; Gobetto, R.; Pellergrino, L.; Milone, L.; Venturello, P. Mechanically induced phase change in Barbituric acid. *Cryst. Growth Des.* **2008**, *8*, 1454–1457. [CrossRef]

45. Schmidt, M.U.; Brning, J.; Glinnemann, J.; Htzler, M.W.; Mçrschel, P.; Ivashevskaya, S.N.; van de Streek, J.; Braga, D.; Maini, L.; Chierotti, M.R.; et al. The thermodynamically stable form of solid barbituric acid: The enol tautomer. *Angew. Chem. Int. Ed.* **2011**, *50*, 7924–7926. [CrossRef]
46. Paul, M.E.; da Silva, T.H.; King, M.D. True polymorphic phase transition or dynamic crystal disorder? an investigation into the unusual phase behavior of barbituric acid dihydrate. *Cryst. Growth Des.* **2019**, *19*, 4745–4753. [CrossRef]
47. Braga, D.; Cadoni, M.; Grepioni, F.; Maini, L.; Rubini, K. Gas–solid reactions between the different polymorphic modifications of barbituric acid and amines. *CrystEngComm* **2006**, *8*, 756–763. [CrossRef]

Article

Improving Nefiracetam Dissolution and Solubility Behavior Using a Cocrystallization Approach

Xavier Buol [1,*], Koen Robeyns [1], Camila Caro Garrido [1], Nikolay Tumanov [2], Laurent Collard [1], Johan Wouters [2] and Tom Leyssens [1,*]

[1] Institute of Condensed Matter and Nanosciences, UCLouvain, 1 Place Louis Pasteur, B-1348 Louvain-la-Neuve, Belgium; koen.robeyns@uclouvain.be (K.R.); camila.carogarrido@student.uclouvain.be (C.C.G.); laurent.collard@uclouvain.be (L.C.)

[2] Namur Research Institute for Life Sciences (Narilis), Chemistry Department, UNamur, 61 rue de Bruxelles, B-5000 Namur, Belgium; nikolay.tumanov@unamur.be (N.T.); johan.wouters@unamur.be (J.W.)

* Correspondence: xavier.buol@uclouvain.be (X.B.); tom.leyssens@uclouvain.be (T.L.); Tel.: +32-047-592-5798 (X.B.); +32-10-472-711 (T.L.)

Received: 17 June 2020; Accepted: 7 July 2020; Published: 9 July 2020

Abstract: In this work, we are the first to identify thirteen cocrystals of Nefiracetam, a poor water-soluble nootropic compound. Three of which were obtained with the biocompatible cocrystallization agents citric acid, oxalic acid, and zinc chloride. These latter have been fully structurally and physically characterized and the solubility, dissolution rate, and stability were compared to that of the initial Active Pharmaceutical Ingredient (API).

Keywords: nefiracetam; solid state; cocrystals; solubility; dissolution rate; stability; formulation

---

1. Introduction

Nefiracetam, a member of the racetam family, is a nootropic compound typically administered as a cognitive enhancer. This API ((N-(2,6-dimethylphenyl)-2-(2-oxopyrrolidin-1-yl)acetamide)) known under the label DM-9384 is a pyrrolidone derivative produced and developed during the 90s by Daiichi Seiyaku and marketed in 2002 in Japan [1,2]. Recently, a polymorph screen by our group [3] revealed the existence of three anhydrous polymorphs as well as a monohydrate form with the asymmetric units illustrated in Figure 1. Different solid-state forms exhibit distinct physicochemical properties such as the melting point, hygroscopicity, solubility, dissolution rate, and bioavailability. Polymorphs FI and FII are enantiotropically related anhydrous forms with FI found to be the most stable form below 140 °C, whereas FII is stable above this temperature. Polymorph FIII is metastable at all temperatures. Evaluation of the solubility/dissolution rate differences between polymorphs was challenging due to the fast solvent-mediated transformation of the meta-stable forms into the most stable one in suspension.

Therefore, as polymorphism does not seem a reasonable approach to modulate the dissolution/solubility properties of this drug, due to the high-energy state of its polymorphs, we here investigate an alternative approach focusing on cocrystal and ionic cocrystal formation [4–8]. Such an approach seems promising for Nefiracetam as it is a non-ionizable, low-water soluble drug (Biopharmaceutical Classification System (BCS) II) displaying synthons favorable for hydrogen bonding and metal coordination, but rendering salt-formation impossible. The FDA guidance defines cocrystals as "solids that are crystalline materials composed of two or more different molecules generally held together by hydrogen bonds in the same crystal lattice" [9,10]. Considering that the molecular components have to be neutral and solid under ambient conditions, salts and solvates can be excluded from this terminology [9,11]. Ionic cocrystals (ICC) can also be included in this category with the specificity that they result from cocrystallization between a neutral organic molecule and an inorganic salt through complexation [12,13]. In the last two decades, cocrystallization has been widely used on pharmaceutical compounds to tune the solubility and/or dissolution rate [14–17], offering novel patentability opportunities in parallel [18]. Metaxalone cocrystals [19] with succinic acid, adipic acid, fumaric acid, salycilic acid, and maleic acid were patented in 2014 showing enhanced physicochemical properties. In 2006, Mc Amara et al. [20] showed that the use of glutaric acid as a cocrystallization agent increased the water solubility of a non-disclosed API about eighteen times and the dissolution rate about three times. Brittain et al. [21–23] recently reported the increasing importance of pharmaceutical cocrystals in the last decade.

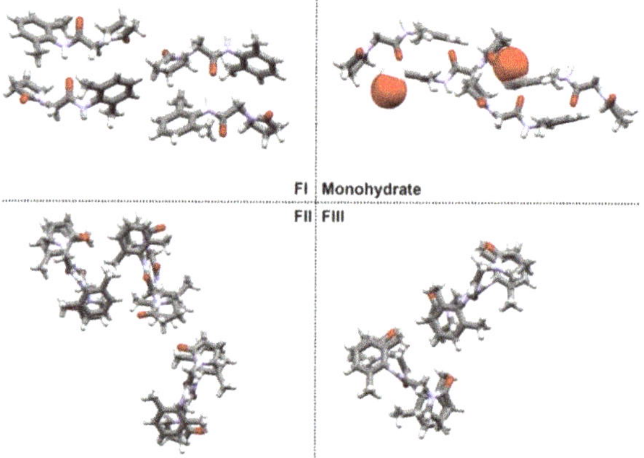

Figure 1. View of the crystal packing along the a-axis of each (pseudo)-polymorphic form of Nefiracetam (*N*-(2,6-dimethylphenyl)-2-(2-oxopyrrolidin-1-yl)acetamide). The spacefill representation is used to highlight water molecules in the monohydrate crystal structure.

2. Materials and Methods

Starting Materials. Nefiracetam was purchased from Xiamen Top Health (Fujian, China). The solvents were sourced from VWR (Leuven, Belgium) and directly used without any purification steps. In the cases of ethyl acetate and acetonitrile, they were dried using a dessicant as the calcium hydride (CaH_2). Nefiracetam was purified by a slurry crystallization in ethyl acetate (room temperature, ca. 300 rpm and overnight) and the solid phase was filtrated, washed, and dried. All coformers were commercially available from Alfa Aesar (Kandel, Germany), Acros Organics (Leuven, Belgium), TCI (Zwijndrecht, Belgium), and Merck (Overijse, Belgium) and used as received.

Nefiracetam Cocrystals Screening. A total of 133 cocrystallization agents typically used in cocrystallization efforts were selected among carboxylic acids, amides, sugars, inorganic salts and amino acids, other racetams, and profens. The cocrystal screening was performed through liquid assisted grinding (LAG) using a MM 400 Mixer Mill grinder manufactured by Retsch (Haan, Germany). The device is equipped with two grinding cells in which five 2 mL Eppendorf tubes can be set. To do so, equimolar amounts of Nefiracetam (0.2 mmol) and coformer (0.2 mmol) were weighted in an Eppendorf, and 4–5 stainless steel beads and 10 µL of solvent (methanol) were appended. Once the jars were filled, the milling runs for 90 min at 30 Hz.

For the three cocrystals studied in more detail in this work, these experiments were repeated using 12 different solvents in the LAG experiments as the nature of the solvent has been shown to impact on the outcome of such an experiment [24,25]. Using a 2:1 Nefiracetam/acid ratio for oxalic and citric acid and a 1:1 ratio for $ZnCl_2$. The tested solvents were ethanol (EtOH), methanol, acetonitrile (MeCN), tetrahydrofuran, acetone, dichloromethane, chloroform, ethyl acetate, methyl acetate, diethylether, 2-propanol, and water, using approximately 10 µL of solvent.

Single Crystal Growth. Single crystals were mainly obtained from evaporative experiments. Nefiracetam and coformer were added in the molar ratio found in the cocrystal, and solids dissolved by the addition of a sufficient amount of solvent. Solutions were then left to evaporate slowly (over periods ranging from three to seven days) at room temperature, and single crystals retrieved. A 2:1 Nefiracetam-citric acid cocrystal Form I (NCA) and 1:1:1 Nefiracetam-zinc chloride-water ionic cocrystal hydrate (NZCW) suitable crystals were obtained from ethyl acetate, while the 2:1 Nefiracetam-oxalic acid cocrystal (NOA) crystals was from ethyl acetate and tetrahydrofuran. Cooling experiments were performed by preparing supersaturated Nefiracetam and coformer solutions. An excess of Nefiracetam and coformer (equimolar) was added to a given solvent volume at room temperature, and the vials were placed at 15 °C below the related solvent boiling point until the full dissolution. Once the full dissolution was achieved, they were stored at −15 or 9 °C in the case of water. Solid phases were then retrieved and analyzed. The 1:1 Nefiracetam-zinc chloride ionic cocrystal (NZC) suitable crystals were obtained from a cooling crystallization in dried acetonitrile.

Nefiracetam Cocrystal Bulk Material Preparation. The bulk material for dissolution testing was prepared through crystallization from the solution. To identify an appropriate solvent, slurrying experiments were performed by placing excess amounts of Nefiracetam-coformer in suspension in different solvents at 25 °C. Vials were sealed and the suspension was left over three days at 25 °C stirring at 700 rpm using a Cooling Thermomixer HLC manufactured by Ditabis (Pforzheim, Germany). Each vial was seeded with all possible solid forms (Nefiracetam and cocrystal) after 2 h of stirring. After three days, solid phases were retrieved and analyzed. The 12 aforementioned solvents were used for this section. A solvent was then selected in which the system behaves congruently (meaning a full transformation to cocrystal occurred over the three days), and an upscaled solvent-mediated cocrystal formation was performed. This upscaling was performed in an EasyMax 102 (Mettler Toledo, Zaventem, Belgium) crystallizer equipped with a hermetically closed 100 mL flask, under mechanic stirring (150 rpm) at 25 °C. After three days, the solids were retrieved, washed, and dried overnight at 50 °C. The solids were then used for dissolution measurements. NCA was upscaled to 30 g of the bulk material in ethyl acetate while NOA and NZC were upscaled in acetonitrile.

Single-Crystal X-Ray diffraction (SCXRD). Suitable crystals of NZCW to perform the SCXRD analysis have been analyzed using a Rigaku (Neu-Isenburg, Germany) Ultra X18S rotating anode, FOX3D mirrors. The diffracted beams were collected on a MAR345 image plate detector using MoKα (λ = 0.71073 Å). Single-crystal X-ray diffraction data for NCA, 2:1 Nefiracetam-citric acid cocrystal Form II (NCA1), and NZC (100 K) were collected on an Oxford Diffraction Gemini R Ultra diffractometer (Ruby CCD detector using CuKα radiation) and crystals of NOA were measured at the SNBL beamline (Pilatus 2 M hybrid pixel detector), ESRF Grenoble. All the methodologies for the data reduction [26], resolution and refinement [27], and validation [28] were specified as in our previous work [3]. The images of the crystal structures were drawn using the software Mercury 4.1.3 [29]. CCDC 2010261-2010276 contain the supplementary crystallographic data for this paper. These data can be obtained free of charge from The Cambridge Crystallographic Data Centre via www.ccdc.cam.ac.uk/structures.

Powder X-Ray Diffraction (XRPD) data were collected on a diffractometer Bragg-Brentano manufactured by PANalytical (Eindhoven, Netherlands). The X-Ray source, a Ni-filtered CuKα (λ = 1.54179 Å), was used at 40 kV and 30 mA. The detection was carried out using a X'Celerator detector. All the powders were analyzed in a 2θ angle range from 4 to 40° for a total scan time of 6 min 42 s (step size = ca. 0.0167°).

Differential Scanning Calorimetry (DSC) measurements were performed from 25 to 175 °C at a scanning rate of 5 °C·min^{-1} on a TA instrument DSC2500 (Zellik, Belgium) in the cases of NCA and NOA. The temperature range was extended to 275 °C for NZC and NZCW. Solid samples (6–7 mg) were placed in an aluminum crucible (40 µL) with pierced sealed lids and nitrogen was used as purge gas with a flow rate of 50 mL·min^{-1}. Indium was used as a reference. Regarding the DSC measurements performed on the suspected and other confirmed cocrystals, the temperature range was extended to 200 and 225 °C.

Thermogravimetric Analysis (TGA). These analyses were carried out on a TGA-STDA 851e manufactured by Mettler Toledo (Columbus, OH, USA). The samples were analyzed in a temperature range from 25 to 400 °C (600 °C for NZC and NZCW). The scanning rate applied during the analysis was 10 °C·min^{-1}. About 7–10 mg of solid materials were placed in an aluminum oxide crucible and a purge gas (nitrogen) was used with a flow rate of 50 mL·min^{-1}.

Dynamic Vapor Sorption (DVS) analyses were performed at 25 °C on a Q5000 SA from TA instruments (New Castle, DE, USA). Solid samples (weight from 7 to 14 mg) were placed in a hanging platinum crucible and an empty crucible is used as the reference. These crucibles were placed in a chamber with controlled humidity and temperature. Nitrogen was used as flowing gas. Samples were first dried at 40 °C for 1 h before being exposed to variable relative humidity within a range of 10% to 90%. The equilibrium was considered to have been reached when the weight change was less than 0.010 mg·min^{-1} or no weight change has occurred for 2 h.

Moisture Exposure. Samples (100 mg) of NCA, NOA, and NZC were stored for one month at room temperature in a sealed box containing an open water flask. The atmosphere was assumed to be saturated in water (100% RH). Both the XRPD and DSC analyses were performed on those powders after the exposure.

High Performance Liquid Chromatography (HPLC). The calibration line (linear equation) for the Nefiracetam dosage was drawn by interrelating the area under the curve (AUC) with the concentration prepared. To do so, Nefiracetam samples were diluted 1000 times using a 1:1 acetonitrile-Milli-Q water diluent (in volume). Nefiracetam chromatograms were then recorded according to the HPLC parameters: Device: Waters Alliance 2695 (Zellik, Belgium); Column: Waters Sunfire C18, 4.6 × 100 mm, 3.5 μm; Detector: PDA 2998 (extraction at λ = 210 nm); T° = 40 °C; injection volume: 10 μL; flow: 1.2 mL/min; mobile phase A: H_2O + 0.1% H_3PO_4; mobile phase B: CH_3CN + 0.1% H_3PO_4; gradient: 0 to 0.5 min at 30% B; 0.5 to 4.5 min 30% B→90% B; 4.5 to 6.5 min at 90% B. The calibration line is reported in the Supplementary Materials.

UltraViolet Spectroscopy (UVS). The calibration line for UV spectroscopy was drawn in order to dose the Nefiracetam by matching the absorbance at λ_{max} (= 263 nm) and the concentration prepared with a linear equation. Acetonitrile was used as a diluent and blank. Sample UV-absorption spectra were recorded from 300 to 200 nm using a UV-1700 PharmaSpec spectrophotometer manufactured by SHIMADZU (Wemmel, Belgium).

Dissolution Experiments. Nefiracetam FI, NCA, NOA, and NZC were first ground using a mortar and pestle to get approximately the same particle size range. An excess amount of compound was added to 50 mL of solvent at 18 °C in a 100 mL flask under magnetic 100 rpm stirring. The 10 μL sampling occurred using a syringe equipped with a micro filter (0.02 μm). Sampling was performed every 10–15 s during the first minute up to t = 1 min, every 30 s up to t = 5 min, and every minute up to t = 15 min. A final sampling (triplicate) was performed after 24 h considering the equilibrium has been reached. Samples were diluted from 2000 to 5000 times depending on the fraction considered with a MeCN-water Milli-Q (1:1 in volume) diluent. Those fractions were then injected in HPLC and dosed as mentioned above. Concerning NZC, the 10 μL-fractions were diluted 250 times with MeCN before being dosed using a UV-1700 PharmaSpec (SHIMADZU) spectrophotometer. HPLC data as well as the UVS data related to the dissolution experiments are presented in the Supplementary Materials.

3. Results and Discussion

3.1. Cocrystal Screening

A large cocrystal screen has been performed on Nefiracetam using liquid-assisted grinding (LAG) and 133 different coformers. A full list is shown in the Supplementary Materials. Coformers have been selected based on crystal engineering principles, with all presenting synthons prone to cocrystal formation [30,31]. Seventeen cocrystals have been suspected based on XRPD data only, corresponding respectively to a success rate of 13%, which is in alignment with cocrystal success rates as reported in the literature [32]. Here, we report only those cases for which single crystals were obtained. This was the case for the following 13 coformers: 5-hydroxyisophthalic acid, 5-nitroisophthalic acid, 5-cyano-1,3-benzenedicarboxylic acid, 5-bromoisophthalic acid, 2-benzoylbenzoic acid, parahydroxybenzoic acid, (RS)-phenylsuccinic acid, (RS)-2-phenylbutyric acid, (RS)-3-phenyllactic acid, gallic acid, citric acid, oxalic acid, and zinc chloride (Figure 2). Cocrystals are also suspected by XRPD and DSC for the positional coformers 2,4-dihydroxybenzoic acid (β-resorcylic acid), 2,5-hydroxybenzoic acid (gentisic acid) and 3,4-hydroxybenzoic acid (protocatechic acid), and dipiconilic acid but no single crystal confirmation has been obtained so far.

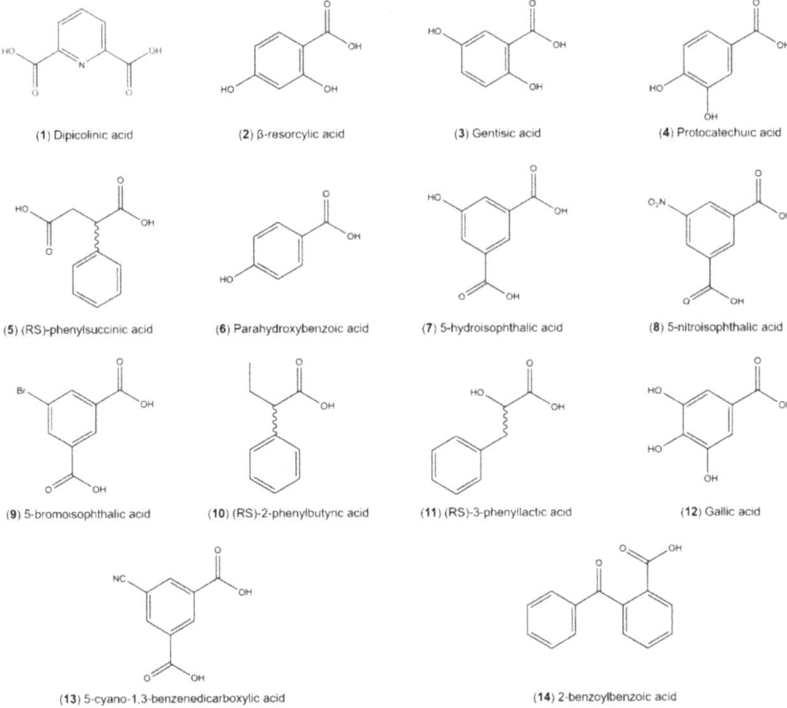

Figure 2. (1–4) Coformers for which a new Nefiracetam cocrystal is suspected and (7–14) coformers for which a single crystal cocrystal confirmation has been achieved.

In a more general manner, successful coformers usually tend to have at least one carboxylic acid group and a phenyl ring present. A similar trend has been observed for the cocrystallization of other racetam compounds [33,34]. Recently, we have also shown inorganic salts to be good cocrystal formers for racetam compounds [35,36], which is also confirmed here. Table 1 summarizes the main results for the 13 coformers leading to confirmed cocrystals. When the ground pattern does not match the simulated pattern from the single crystal, cocrystal polymorphs, solvates, or stoichiometrically diverse cocrystals are suspected [37,38]. This table also summarizes melting/dehydration temperatures of the Nefiracetam forms as well as the obtained cocrystals. Interestingly, we show that a very wide range of properties may be obtained using cocrystallization as a tool. Melting points vary from 68 to 243 °C merely by playing on the nature of the coformer. This shows the potential for variability one can achieve using cocrystallization as a crystal engineering tool. Furthermore, Table 1 also highlights the importance of screening for the various solid-state forms of a given API/coformer system, as cocrystal polymorphs or cocrystal hydrates can be readily encountered.

Table 1. Cocrystal forms identified, together with the melting point. Pharmaceutical systems of interest are highlighted in bold. RT: Room temperature; HT: High temperature.

Coformer	Multiple Cocrystal Forms	Form	Cocrystal/Coformer Melting Point (°C)
None [3] (Nefiracetam (N)/Figure 1)	/	Form I Form II Form III Monohydrate	140 150 n/a 80 (dehydration)
5-nitroisophthalic acid	Yes	1:1 Cocrystal Form I 1:1 Cocrystal Form II	171/259 199
5-hydroxyisopthalic acid	No	1:1 Cocrystal	196/298
5-cyano-1,3-benzenedicarboxylic acid	Yes	1:1 Cocrystal Form I (RT) 1:1 Cocrystal Form II (HT)	160/248 180
5-bromoisophthalic acid	No	1:1 Cocrystal	199/270
(RS)-phenylsuccinic acid	No	2:1 Cocrystal (racemic)	95/166
4-hydroxybenzoic acid	No	1:1 Cocrystal	122/213
(RS)-2-phenylbutyric acid	No	Form I (solid solution)	72/39
(RS)-3-phenyllactic acid	No	1:1 Cocrystal (racemic)	68/95
2-benzoylbenzoic acid	No	1:1 Cocrystal	92/126
Gallic acid	No	4:1:1 Cocrystal hydrate	n/a/251
Citric acid (CA)	Yes	2:1 Cocrystal Form I 2:1 Cocrystal Form II	81 (phase transition) n/a
Oxalic acid (OA)	No	2:1 Cocrystal	162
Zinc chloride (ZC)	Yes	Ionic cocrystal hydrate Ionic cocrystal	210 (dehydration)/290 243

The cocrystals of citric acid, oxalic acid, and zinc chloride are described in more detail as these cocrystals are pharmaceutically acceptable (Figure 3). For the analysis of the others, we refer to the Supplementary Materials. For citric acid (NCA), oxalic acid (NOA), and zinc chloride (**NZC**), LAG experiments all yielded the same outcome, irrespective of the solvent used. For these systems, a more profound form screening was performed using evaporative, slurrying, and cooling crystallization when applicable. Twelve solvents were used for a total of 125 experiments with results shown in the Supplementary Materials. These results led to no additional crystalline form for the NOA cocrystal but did lead to the identification of a second 2:1 cocrystal form for the NCA system and a cocrystal hydrate for the NZC system, highlighting the importance of screening for multiple cocrystal forms once a positive coformer has been identified.

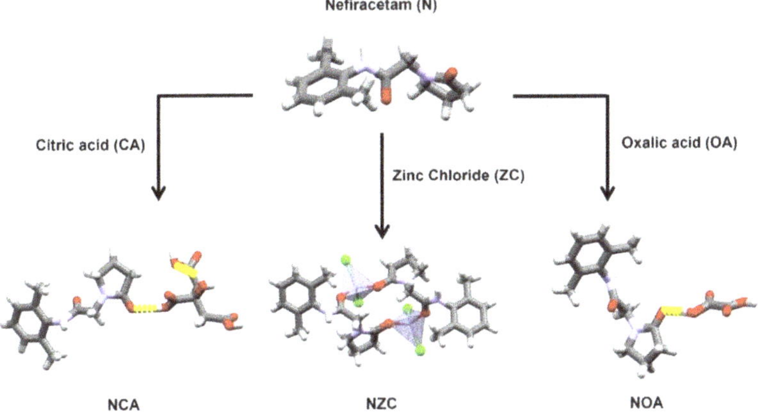

Figure 3. Biocompatible Nefiracetam cocrystals identified and studied in detail in this work. Relevant intramolecular interactions between Nefiracetam and the coformers are highlighted by yellow stick contacts.

3.2. The 2:1 Nefiracetam-Oxalic Acid Cocrystal (NOA)

As mentioned above, only one form of the NOA cocrystal has been identified, which is easily identifiable through its 2θ peaks at 11.3, 13.6, and 15.9°. The bulk product pattern is furthermore shown to match the one simulated from the powder resolution analysis (Figure 4a), which shows NOA to crystallize in the monoclinic $P2_1$ space group. Figure 5a shows a main hydrogen bond C_1^1 (4) chain pattern according to Etter's graph-set notation [39,40] built through N-H···O (2.86 (1) Å, 152.0° as a hydrogen bond distance N···O and NHO angle) hydrogen bonds between Nefiracetam amide moieties. Nefiracetam is linked to the oxalic acid through a hydrogen bonding pattern between the hydroxyl group from both the carboxylic acid moieties and the oxygen of the γ-lactam moiety of two different Nefiracetam molecules. These interactions (2.54 (1) Å, 167 and 171.5°) result in a D_1^1 (2) hydrogen-bond pattern. This assembly creates an overall pattern with Nefiracetam chains along the *a*-axis and oxalic acid acting as a linker between chains (zig-zag feature in Figure 5a represented by a full black line). The main crystallographic parameters related to each form and the refinement parameters are summarized in Table S2 in the Supplementary Materials. Upon heating, NOA shows a sharp melting endotherm at 162 °C (Figure 6c) and right upon melting, a weight loss of about 15% is observed (Figure 6a), corresponding to oxalic acid sublimation [41,42]. After 200 °C, degradation of Nefiracetam is observed. After a prolonged (one-month) exposure under a humidity saturated atmosphere (100% RH), no transformation of NOA into another form or deliquescence were observed by XRPD and DSC measurements (Supplementary Materials).

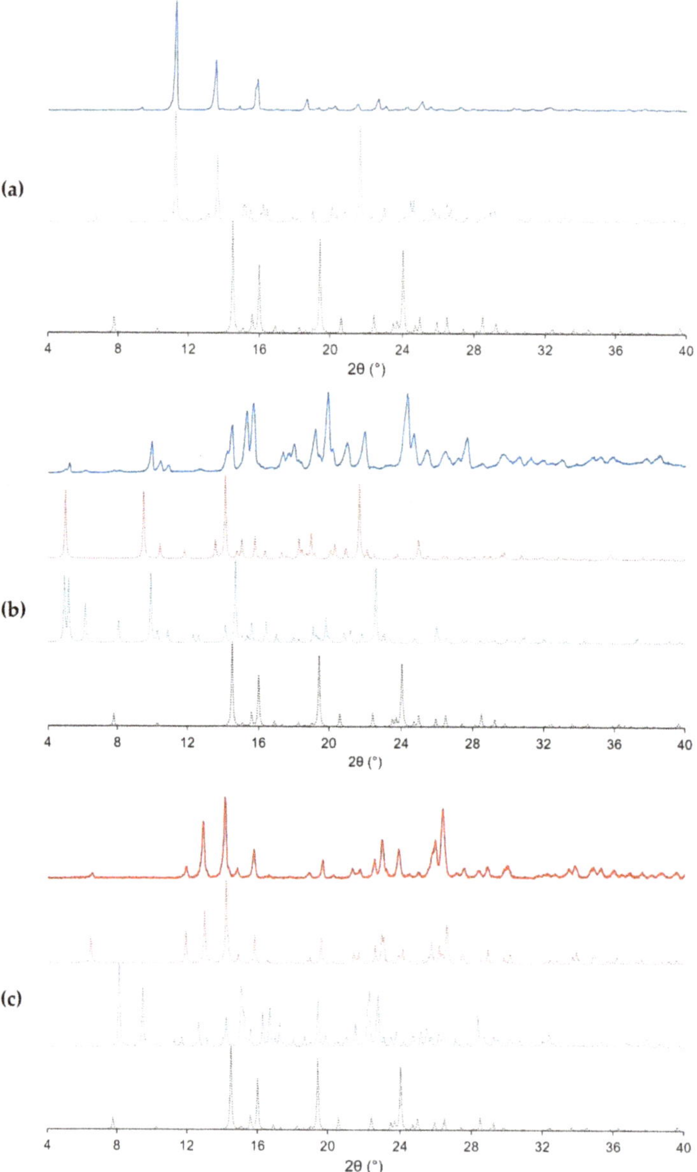

Figure 4. Simulated diffraction patterns (dashed line) and experimental powders (full line) of (**a**) Nefiracetam-oxalic acid cocrystal (NOA) (blue), (**b**) Nefiracetam-citric acid cocrystal (NCA) (blue) and NCA1 (red), and (**c**) Nefiracetam-zinc chloride ionic cocrystal (NZCW) (blue) and NZC (red). Patterns are compared to the Nefiracetam one (black).

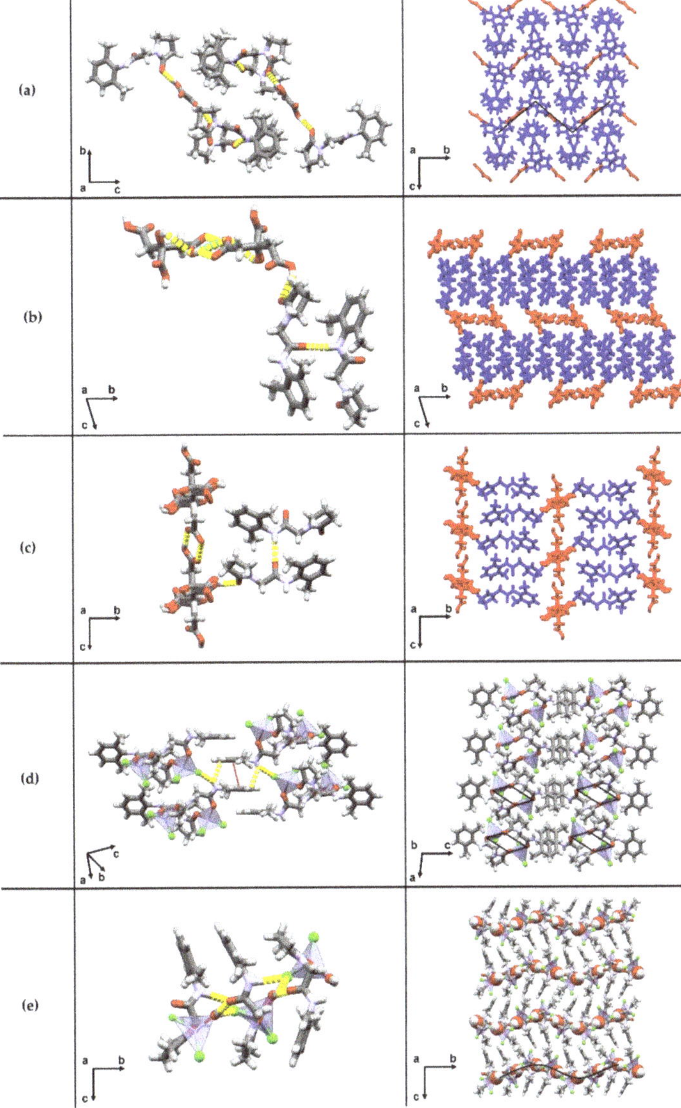

Figure 5. Main intermolecular interactions and crystal packing along the a- and b-axis in the crystal structure of NCA (**a**), NCA1 (**b**), NOA (**c**), NZC (**d**), and NZCW (**e**). The tetragonal geometry of the Zn^{2+} based-complex and the water molecules are respectively highlighted using the polyhedral and spacefill representation. Intermolecular contacts are shown using yellow dashed lines. The red dashed line is used to show the centroid-centroid distance in the NZC crystal structure.

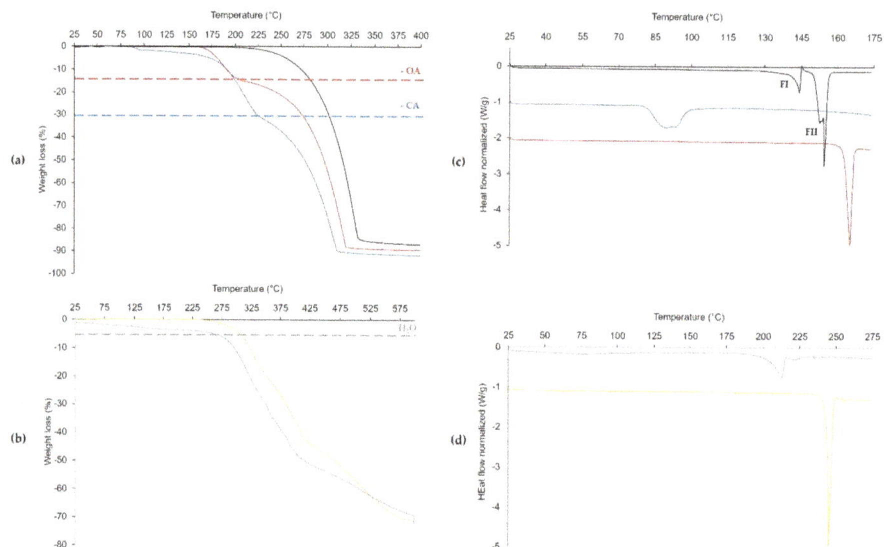

Figure 6. TGA curves (**a,b**) and DSC curves (**c,d**) of Nefiracetam (black), NCA (blue), NOA (red), NZC (yellow), and NZCW (grey).

3.3. The 2:1 Nefiracetam-Citric Acid Cocrystal (NCA and NCA1)

Two different polymorphs of the 2:1 Nefiracetam-citric acid cocrystal have been obtained. All screening experiments led to the same 2:1 cocrystal form (NCA). Single crystals of this form were obtained from a slow evaporation from ethyl acetate. Unexpectedly, the SCXRD diffraction experiment of NCA revealed also the presence of a second 2:1 form NCA1, in the same crystal. This latter is likely due to a phenomena of cross-nucleation, i.e., the heterogeneous nucleation of Form II (NCA1, daughter form) on the surface of the other (NCA, parent form) [43,44]. This Form II was never observed in sufficient quantity in any of the screening or bulk material preparation experiments, and is therefore likely a kinetic form, crystallizing under high supersaturated conditions, which occurs at the end of the solvent evaporation process. As during upscaling, the crystallization solvent is never fully evaporated, and the material is washed, this polymorph is very likely not present in the bulk material that fits the NCA simulated powder pattern in Figure 4b. NCA crystallizes in the triclinic P-1 space group. As for the anhydrous forms of Nefiracetam [3], the Nefiracetam molecules are stacked in a chain, by intermolecular amide-amide hydrogen bonds (N-H···O, 2.773 (5) Å, 168.2°) leading to a C_1^1 (4) hydrogen-bond pattern. In Figure 5b, the recognition between Nefiracetam and citric acid is due to two extra D_1^1 (2) intermolecular hydrogen bonds (O-H···O, 3.18 (1) Å, 146.64° and O-H···O, 2.596 (5) Å, 168.01°) which link one citric acid molecule to two Nefiracetam molecules through the oxygen of the γ-lactam moiety (Nefiracetam) and the hydroxyl of a carboxylic acid (citric acid). Moreover, a ring feature (R_2^2 (8) with 2.654 (6) Å and 152.45°) leading to a cyclic citric acid dimer is observable between two carboxylic acids from different citric acid molecules. Finally, an intramolecular hydrogen bond O-H···O (S_1^1 (3), 3.176 (7) Å, 121.48°) is present in each citric acid molecule.

Polymorph NCA1 shows disorder and crystallizes in the monoclinic $P2_1/c$ space group. Intermolecular interactions between citric acid and Nefiracetam are identical to those mentioned for the first NCA form (Figure 5c). In Figure 6a, TGA data shows three respective weight losses in the case of NCA. A first weight loss occurs around 80 °C, and must likely correspond to the loss of water molecules trapped in the channel-like structure observed for this cocrystal, evidencing that NCA is likely a non-stoichiometric hydrate. The following waves correspond to citric acid degradation (±28%, 175 °C) [45] and Nefiracetam degradation (from 225 °C), respectively. The DSC endotherm at around

80 °C corresponds to a potential dehydration of a non-stoichiometric hydrate, but could potentially also correspond to a melting point. DVS shows a deliquescent (+20% in weight) behavior at RH > 80%. When exposing NCA to a saturated humidity atmosphere for a long time period (one month), deliquescence is followed by Nefiracetam monohydrate crystallization (Supplementary Materials).

3.4. The 1:1 Nefiracetam-Zinc Chloride Ionic Cocrystal (**NZC**)

The form screening with zinc chloride led to two different cocrystals. The full screening results are presented in the Supplementary Materials. Slurrying crystallization in a congruent and dried solvent, led to a 1:1 ionic cocrystal (NZC) wherein Nefiracetam is coordinated with the Zn^{2+}. In addition, evaporative experiments at room temperature with water traces in the solvent (ethyl acetate) led to a 1:1:1 hydrated ionic cocrystal (so called NZCW) form whose structure was solved showing a zinc chloride complex with Nefiracetam and one water molecule. Only NZC was upscaled in dried acetonitrile. We were unsuccessful at obtaining larger amounts of NZCW as all experimental conditions tried led to mixtures of the Nefiracetam monohydrate, NZCW, and anhydrous NZC. The experimental and simulated XRPD data related to each system are presented in Figure 4c. Single crystals obtained from slowly cooling a supersaturated acetonitrile solution, yielded suitable monoclinic $C2/c$ cube-like single crystals of the anhydrate (NZC) wherein Nefiracetam is complexed to zinc chloride showing a tetragonal geometry around the Zn^{2+} ion. Nefiracetam molecules are bound to each other by two identical (by symmetry), tetrahedral complexes as shown in Figures 2 and 5d (right). Complexation occurs between zinc chloride and the γ-lactam C=O moiety of a first molecule and the amide C=O from a second molecule and vice versa. The donor moiety (N-H) from the amide is involved in a hydrogen bond pattern ($C_1^1(4)$, 3.496(1) Å, 146.49°) with the chloride from the tetragonal complex along the a-axis. These tetrahedral complexes form a network of Nefiracetam dimer molecules as shown with black diamonds in Figure 5d. The π-stacking interactions are present between the Nefiracetam aromatic moieties with a centroid-centroid distance of about 3.54°. This stacking may explain the higher density (1.637 Mg/m^3) comparing NZC with NZCW.

NZCW crystallizes in the monoclinic $P2_1$ space group. The amide-amide hydrogen bond between Nefiracetam molecules as described above remains present but does not lead to a C_1^1 (4) chain pattern as in the previous cases. Indeed, the Nefiracetam chains are interrupted by water molecules and Nefiracetam amide moieties are now involved in three D_1^1 (2) hydrogen-bond patterns with respectively a water hydrogen (O-H···H, 2.734 (2) Å, 174.33°), a second Nefiracetam amide moiety (N-H···O, 3.019 (3) Å, 164.80°), and with a chloride ion (N-H···Cl, 3.331 (2) Å, 169.63°). An infinite chain (O-H···Cl, C_1^1 (4), 3.154 (2) Å, 169.19°) is also observed involving the second hydrogen from the water molecule and a chloride ion. In the presence of zinc chloride, the carbonyl not belonging to the pyrrolidone group is coordinated to the Zn^{2+} cation as well as to the water oxygen atom leading to a tetrahedral complex. This complex is highlighted in Figure 5e and forms "wave" chains stacking Nefiracetam densely (1.514 Mg/m^3) along the b-axis. NZC shows a single melting endotherm at 243 °C. On the other hand, NZCW shows a continuous dehydration behavior (coordinated-water loss of ±4.5% corresponds to 1 equivalent of water) up until 175 °C. The DSC analysis then shows an endothermic peak at 200 °C which potentially corresponds to the melting of another polymorphic form of the ionic cocrystal or is a mere effect of the water still being present in the DSC capsule. To observe the potential transformation of NZC into NZCW, NZC was stored for one month in a saturated humidity chamber at room temperature. Even though, the DSC analysis indicates traces of the hydrated form (small endotherm around 210 °C), XRPD shows no such transformation. If NZCW is present in the bulk, it would only be in trace amount (Supplementary Materials). Therefore, NZC could be a potentially interesting form to market.

3.5. Solubility and Dissolution Profile of Nefiracetam Cocrystals

Ethanol and acetonitrile were selected to study the dissolution profile at 18 °C (100 rpm) of NCA, NOA, and NZC in comparison to the API (Nefiracetam FI). As the cocrystals considered here behave incongruently in water, with Nefiracetam monohydrate crystallizing out rapidly, we were unable to determine the apparent cocrystal solubility in this solvent. Therefore, we decided to perform the dissolution studies in EtOH and MeCN as we aim at underlining the potential impact cocrystals can have on physicochemical parameters. NCA, NOA, and NZC are incongruent in EtOH leading to a final slurry of Nefiracetam, whereas a congruent dissolution is observed in MeCN. In the case of an incongruent system in EtOH, a "spring-parachute" [46,47] behavior (Figure 7a) is expected meaning that the concentration reaches a maximum ("apparent solubility" [48,49]) before dropping to the solubility of the drug form crystallizing out (here Nefiracetam FI). The final solubility of Nefiracetam might, however, still be different as solution interactions will occur. In the present case, the cocrystal dissolution kinetics combined to the crystallization kinetics of Nefiracetam FI occur too rapidly for the "spring-parachute" behavior to be observable. Nevertheless, one notices a clear impact of the coformer on the solubility of Nefiracetam (Table 2) with the overall solubility at 18 °C observed in both cases around 750 mM (+40%) instead of 519 mM (Figure 7c) in EtOH. The presence of $ZnCl_2$ on the other hand, reduces solubility to about 199 mM.

Table 2. Dissolution rates and solubility at 18 °C in EtOH in the case of Nefiracetam FI, NCA, NOA, and NZC.

Solid Forms	Dissolution Rate * (mM/s)/Time (min)	Solubility (mM, 18 °C)
Nefiracetam FI	1299/0.2	519.59 (±11.88)
NCA	419/0.9	753.47 (±25.65)
NOA	246/1.5	737.53 (±8.58)
NZC	332/0.3	199.46 (±1.61)

* Dissolution rates correspond to the rate calculated when half the solubility is reached, with the corresponding time also given.

Table 3 also shows a true cocrystal solubility in MeCN where all cocrystals behave congruently. Here, one also clearly sees the importance of the coformer nature on the overall solubility, with citric acid slightly increasing the amount of Nefiracetam dissolved, whereas oxalic acid reduces this amount by about 30% in comparison to Nefiracetam FI. A striking drop in Nefiracetam present in the solution of about 90% is observed using the NZC cocrystal.

Table 3. Dissolution rates and solubility at 18 °C in MeCN in the case of Nefiracetam FI, NCA, NOA, and NZC.

Solid Forms	Dissolution Rate * (mM/s)/Time (min)	Solubility (mM, 18 °C)	Solubility Product (Ks, 18 °C)	pKs
Nefiracetam FI	969/0.1	290.83 (±10.97)	/	/
NCA	323/0.5	322.87 (±4.32)	1.35×10^{-1}	0.87
NOA	538/0.2	215.03 (±1.03)	3.98×10^{-2}	1.40
NZC	64/0.2	25.45 (±0.63)	6.48×10^{-4}	3.19

* Dissolution rates correspond to the rate calculated when half the solubility is reached, with the corresponding time also given.

The changes in solubility also reflect in the dissolution rate. However, they are all less important in comparison with the dissolution rate of the parent compound. It should nevertheless be noticed that all solid forms dissolve very rapidly, with half of the maximum solubility reached in the first minute and full solubility reached in the first 5 min of adding powder to the reactor. Therefore, we expect the bioavailability to be strongly impacted by the type of solid form used, not for the impact on the time required to reach solubility, but rather by the impact on the solubility value.

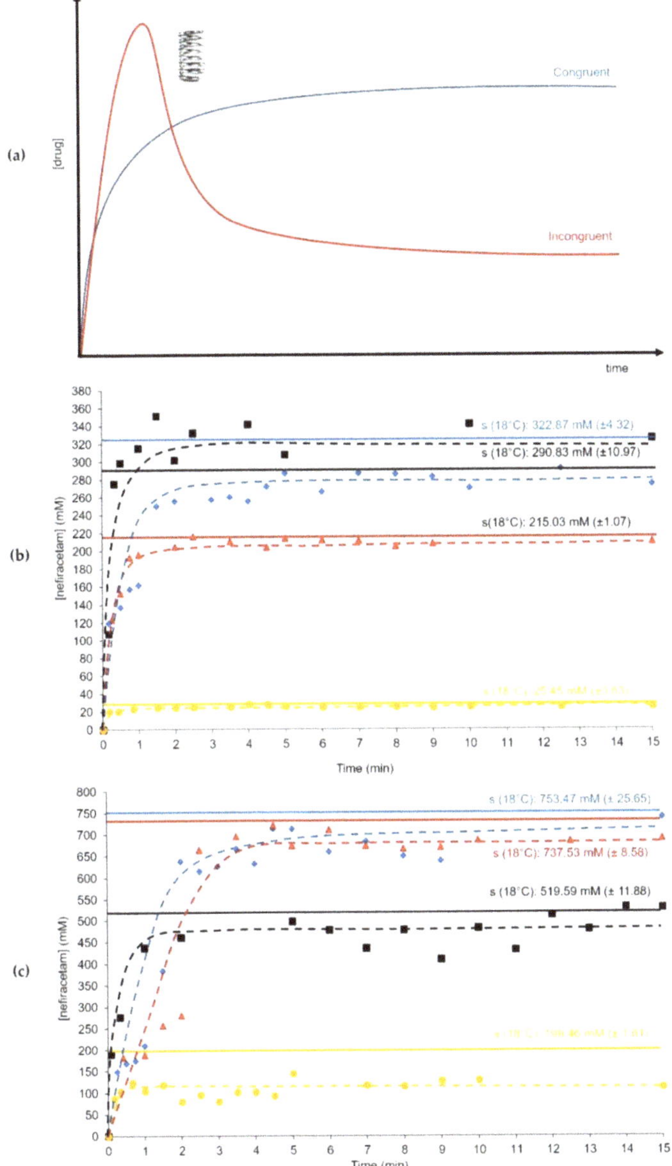

Figure 7. (**a**) Theoretical dissolution curve in the case of a congruent and an incongruent system (with spring-parachute behavior). Experimental dissolution curves of Nefiracetam FI (black ■), NCA (blue ♦), NOA (red ▲), and NZC (yellow •) in (**b**) MeCN at 18 °C with a 100-rpm stirring and (**c**) EtOH at 18 °C with a 100-rpm stirring.

4. Conclusions

Nefiracetam cocrystal screening led to the identification of 13 novel cocrystal systems. Among these, three biocompatible systems were identified (NCA, NOA, and NZC) and their physicochemical properties (stability, solubility, and dissolution) were evaluated and compared to those of the parent compound Nefiracetam. The zinc chloride ionic cocrystal showed an impressive improvement in thermal stability, but strongly reduced the solubility by about 90% in organic solvents in comparison with the parent drug. The use of dicarboxylic acid coformers can induce a complete different behavior, with citric acid improving the solubility, while oxalic acid reduces this latter when the cocrystal behaves congruently. For non-congruent systems, the presence of the coformer always increases the overall solubility. All cocrystals, as well as the parent compound, are rapidly dissolving. Overall, this study confirms that cocrystallization can be used as an effective tool to impact the physico-chemical properties of a drug compound. A relevant choice in the coformer can either help improve formulation stability, or bioavailability of the drug. The coformer selection is of utter importance, as strong variations can occur depending on the nature of the selected coformer.

Supplementary Materials: The aforementioned content is available online at http://www.mdpi.com/1999-4923/12/7/653/s1. The coformer list (133) can be found in the Supplementary Materials. The full NCA, NOA, and NZC form screening and the humidity exposure (DVS) as well are presented in the Supplementary Materials. All the XRPD (experimental and simulated) and DSC data of the other positive hits (suspected and confirmed) mentioned in the manuscript are in the Supplementary Materials too. The crystallographic data and the specific refinement method related to each structure solved in this work are presented in the Supplementary Materials.

Author Contributions: Conceptualization, T.L. and X.B.; methodology, X.B.; software, K.R. and N.T.; validation, X.B.; formal analysis, X.B., K.R., N.T., and L.C.; investigation, X.B. and C.C.G.; resources, T.L. and J.W.; data curation, X.B., K.R., and N.T.; writing—original draft preparation, X.B.; writing—review and editing, X.B., T.L., J.W., and K.R.; visualization, X.B., T.L., J.W., and K.R., X.B.; supervision, T.L. and J.W.; project administration, T.L.; funding acquisition, T.L. All authors have read and agreed to the published version of the manuscript.

Funding: This research was funded by the "Fonds Européen de Développement Régional" and the "Région Wallonne" in the operational framework Wallonie-2020.EU (Tera4all).

Conflicts of Interest: The authors declare no conflict of interest.

References

1. Wang, P.L.; Wang, H.B.; Ding, W.L.; Liu, Z.Q.; Xing, Z.T. N-(2, 6-Dimethylphenyl)-2-(2-oxo-1-pyrrolidin-1-yl) acetamide. *Acta Crystallogr. Sect. E* **2006**, *62*, 3838–3839. [CrossRef]

2. Crespi, F. Nefiracetam. Daiichi Seiyaku. *Curr. Opin. Investig. Drugs* **2002**, *3*, 788–793. [PubMed]

3. Buol, X.; Robeyns, K.; Tumanov, N.; Wouters, J.; Leyssens, T. Identifying, Characterizing, and Understanding Nefiracetam in Its Solid State Forms: A Potential Antidementia Drug. *J. Pharm. Sci.* **2019**, *108*, 3616–3622. [CrossRef]

4. Duggirala, N.K.; Perry, M.L.; Almarsson, Ö.; Zaworotko, M.J. Pharmaceutical cocrystals: Along the path to improved medicines. *Chem. Commun.* **2016**, *52*, 640–655. [CrossRef] [PubMed]

5. Schultheiss, N.; Newman, A. Pharmaceutical cocrystals and their physicochemical properties. *Cryst. Growth Des.* **2009**, *9*, 2950–2967. [CrossRef] [PubMed]

6. Bolla, G.; Nangia, A. Pharmaceutical cocrystals: Walking the talk. *Chem. Commun.* **2016**, *52*, 8342–8360. [CrossRef]

7. Shan, N.; Zaworotko, M.J. The role of cocrystals in pharmaceutical science. *Drug Discov. Today* **2008**, *13*, 440–446. [CrossRef]

8. Grepioni, F.; Wouters, J.; Braga, D.; Nanna, S.; Fours, B.; Coquerel, G.; Longfils, G.; Rome, S.; Aerts, L.; Quéré, L. Ionic co-crystals of racetams: Solid-state properties enhancement of neutral active pharmaceutical ingredients via addition of Mg^{2+} and Ca^{2+} chlorides. *CrystEngComm* **2014**, *16*, 5887–5896. [CrossRef]

9. Aitipamula, S.; Banerjee, R.; Bansal, A.K.; Biradha, K.; Cheney, M.L.; Choudhury, A.R.; Desiraju, G.R.; Dikundwar, A.G.; Dubey, R.; Duggirala, N. Polymorphs, salts, and cocrystals: What's in a name? *Cryst. Growth Des.* **2012**, *12*, 2147–2152. [CrossRef]

10. Thipparaboina, R.; Kumar, D.; Chavan, R.B.; Shastri, N.R. Multidrug co-crystals: Towards the development of effective therapeutic hybrids. *Drug Discov. Today* **2016**, *21*, 481–490. [CrossRef]

11. Bond, A.D. What is a co-crystal? *CrystEngComm* **2007**, *9*, 833–834. [CrossRef]

12. Braga, D.; Grepioni, F.; Lampronti, G.I.; Maini, L.; Turrina, A. Ionic co-crystals of organic molecules with metal halides: A new prospect in the solid formulation of active pharmaceutical ingredients. *Cryst. Growth Des.* **2011**, *11*, 5621–5627. [CrossRef]

13. Braga, D.; Grepioni, F.; Shemchuk, O. Organic–inorganic ionic co-crystals: A new class of multipurpose compounds. *CrystEngComm* **2018**, *20*, 2212–2220. [CrossRef]

14. Singh, D.; Bedi, N.; Tiwary, A.K. Enhancing solubility of poorly aqueous soluble drugs: Critical appraisal of techniques. *J. Pharm. Investig.* **2018**, *48*, 509–526. [CrossRef]

15. Shevchenko, A.; Bimbo, L.M.; Miroshnyk, I.; Haarala, J.; Jelínková, K.; Syrjänen, K.; van Veen, B.; Kiesvaara, J.; Santos, H.A.; Yliruusi, J. A new cocrystal and salts of itraconazole: Comparison of solid-state properties, stability and dissolution behavior. *Int. J. Pharm.* **2012**, *436*, 403–409. [CrossRef] [PubMed]

16. Shan, N.; Perry, M.L.; Weyna, D.R.; Zaworotko, M.J. Impact of pharmaceutical cocrystals: The effects on drug pharmacokinetics. *Expert Opin. Drug Metab. Toxicol.* **2014**, *10*, 1255–1271. [CrossRef] [PubMed]

17. Thakuria, R.; Delori, A.; Jones, W.; Lipert, M.P.; Roy, L.; Rodríguez-Hornedo, N. Pharmaceutical cocrystals and poorly soluble drugs. *Int. J. Pharm.* **2013**, *453*, 101–125. [CrossRef] [PubMed]

18. Leyssens, T.; Harmsen, B. WO2019143829-Fasoracetam Crystalline Forms. 2019. Available online: https://patentscope.wipo.int/search/en/detail.jsf?docId=WO2019143829 (accessed on 26 May 2020).

19. Holland, J.; Frampton, C.; Chorlton, A.; Gooding, D. Metaxalone Cocrystals. U.S. Patent No. 8,871,793, 28 October 2014.

20. McNamara, D.P.; Childs, S.L.; Giordano, J.; Iarriccio, A.; Cassidy, J.; Shet, M.S.; Mannion, R.; O'Donnell, E.; Park, A. Use of a glutaric acid cocrystal to improve oral bioavailability of a low solubility API. *Pharm. Res.* **2006**, *23*, 1888–1897. [CrossRef]

21. Brittain, H.G. Cocrystal Systems of Pharmaceutical Interest: 2012–2014. *Profiles Drug Subst. Excip. Relat. Methodol.* **2019**, *44*, 415–443.

22. Brittain, H.G. Cocrystal systems of pharmaceutical interest: 2010. *Cryst. Growth Des.* **2012**, *12*, 1046–1054. [CrossRef]

23. Brittain, H.G. Cocrystal systems of pharmaceutical interest: 2011. *Cryst. Growth Des.* **2012**, *12*, 5823–5832. [CrossRef]

24. Friščić, T.; Childs, S.L.; Rizvi, S.A.; Jones, W. The role of solvent in mechanochemical and sonochemical cocrystal formation: A solubility-based approach for predicting cocrystallisation outcome. *CrystEngComm* **2009**, *11*, 418–426. [CrossRef]

25. Fischer, F.; Heidrich, A.; Greiser, S.; Benemann, S.; Rademann, K.; Emmerling, F. Polymorphism of mechanochemically synthesized cocrystals: A case study. *Cryst. Growth Des.* **2016**, *16*, 1701–1707. [CrossRef]

26. Rigaku, O.D. *CrysAlisPro Software System*; Version 1.171. 38.41 l; Rigaku Coorporation: Oxford, UK, 2015.

27. Sheldrick, G.M. Crystal structure refinement with SHELXL. *Acta Crystallogr. Sect. C* **2015**, *71*, 3–8. [CrossRef] [PubMed]

28. Spek, A.L. Structure validation in chemical crystallography. *Acta Crystallogr. Sect. D* **2009**, *65*, 148–155. [CrossRef] [PubMed]

29. Macrae, C.F.; Bruno, I.J.; Chisholm, J.A.; Edgington, P.R.; McCabe, P.; Pidcock, E.; Rodriguez-Monge, L.; Taylor, R.; Streek, J.V.D.; Wood, P.A. Mercury CSD 2.0–new features for the visualization and investigation of crystal structures. *J. Appl. Crystallogr.* **2008**, *41*, 466–470. [CrossRef]

30. Blagden, N.; De Matas, M.; Gavan, P.T.; York, P. Crystal engineering of active pharmaceutical ingredients to improve solubility and dissolution rates. *Adv. Drug Deliv. Rev.* **2007**, *59*, 617–630. [CrossRef]

31. Desiraju, G.R. Supramolecular synthons in crystal engineering—A new organic synthesis. *Angew. Chem. Int. Ed. Engl.* **1995**, *34*, 2311–2327. [CrossRef]

32. James, S.L.; Adams, C.J.; Bolm, C.; Braga, D.; Collier, P.; Friščić, T.; Grepioni, F.; Harris, K.D.; Hyett, G.; Jones, W. Mechanochemistry: Opportunities for new and cleaner synthesis. *Chem. Soc. Rev.* **2012**, *41*, 413–447. [CrossRef]

33. Springuel, G.R.; Robeyns, K.; Norberg, B.; Wouters, J.; Leyssens, T. Cocrystal formation between chiral compounds: How cocrystals differ from salts. *Cryst. Growth Des.* **2014**, *14*, 3996–4004. [CrossRef]

34. Springuel, G.R.; Norberg, B.; Robeyns, K.; Wouters, J.; Leyssens, T. Advances in pharmaceutical co-crystal screening: Effective co-crystal screening through structural resemblance. *Cryst. Growth Des.* **2012**, *12*, 475–484. [CrossRef]

35. Song, L.; Shemchuk, O.; Robeyns, K.; Braga, D.; Grepioni, F.; Leyssens, T. Ionic cocrystals of etiracetam and levetiracetam: The importance of chirality for ionic cocrystals. *Cryst. Growth Des.* **2019**, *19*, 2446–2454. [CrossRef]

36. Shemchuk, O.; Song, L.; Tumanov, N.; Wouters, J.; Braga, D.; Grepioni, F.; Leyssens, T. Chiral resolution of RS-oxiracetam upon cocrystallization with pharmaceutically acceptable inorganic salts. *Cryst. Growth Des.* **2020**. [CrossRef]

37. Porter Iii, W.W.; Elie, S.C.; Matzger, A.J. Polymorphism in carbamazepine cocrystals. *Cryst. Growth Des.* **2008**, *8*, 14–16. [CrossRef] [PubMed]

38. Leyssens, T.; Springuel, G.; Montis, R.; Candoni, N.; Veesler, S.P. Importance of solvent selection for stoichiometrically diverse cocrystal systems: Caffeine/maleic acid 1: 1 and 2: 1 cocrystals. *Cryst. Growth Des.* **2012**, *12*, 1520–1530. [CrossRef]

39. Etter, M.C.; MacDonald, J.C.; Bernstein, J. Graph-set analysis of hydrogen-bond patterns in organic crystals. *Acta Crystallogr. Sect. B* **1990**, *46*, 256–262. [CrossRef]

40. Bernstein, J.; Davis, R.E.; Shimoni, L.; Chang, N.L. Patterns in hydrogen bonding: Functionality and graph set analysis in crystals. *Angew. Chem. Int. Ed. Engl.* **1995**, *34*, 1555–1573. [CrossRef]

41. Oxalic Acid-Registration Dossier-ECHA. Available online: https://echa.europa.eu/registration-dossier/-/registered-dossier/14786/4/3 (accessed on 26 May 2020).

42. United States Department of Labor-Occupational Safety and Health Administration-Oxalic Acid. Available online: https://www.osha.gov/chemicaldata/chemResult.html?RecNo=270 (accessed on 26 May 2020).

43. Yu, L. Survival of the fittest polymorph: How fast nucleater can lose to fast grower. *CrystEngComm* **2007**, *9*, 847–851. [CrossRef]

44. Looijmans, S.F.; Cavallo, D.; Yu, L.; Peters, G.W. Cross-nucleation between polymorphs: Quantitative modeling of kinetics and morphology. *Cryst. Growth Des.* **2018**, *18*, 3921–3926. [CrossRef]

45. Barbooti, M.M.; Al-Sammerrai, D.A. Thermal decomposition of citric acid. *Thermochim. Acta* **1986**, *98*, 119–126. [CrossRef]

46. Bavishi, D.D.; Borkhataria, C.H. Spring and parachute: How cocrystals enhance solubility. *Prog. Cryst. Growth Charact. Mater.* **2016**, *62*, 1–8. [CrossRef]

47. Guzmán, H.R.; Tawa, M.; Zhang, Z.; Ratanabanangkoon, P.; Shaw, P.; Gardner, C.R.; Chen, H.; Moreau, J.P.; Almarsson, Ö.; Remenar, J.F. Combined use of crystalline salt forms and precipitation inhibitors to improve oral absorption of celecoxib from solid oral formulations. *J. Pharm. Sci.* **2007**, *96*, 2686–2702. [CrossRef] [PubMed]

48. Nicoud, L.; Licordari, F.; Myerson, A.S. Estimation of the Solubility of Metastable Polymorphs: A Critical Review. *Cryst. Growth Des.* **2018**, *18*, 7228–7237. [CrossRef]

49. Nicoud, L.; Licordari, F.; Myerson, A.S. Polymorph control in batch seeded crystallizers. A case study with paracetamol. *CrystEngComm* **2019**, *21*, 2105–2118. [CrossRef]

 *pharmaceutics*

Article

Dissolution Advantage of Nitazoxanide Cocrystals in the Presence of Cellulosic Polymers

Reynaldo Salas-Zúñiga [1,2], Christian Rodríguez-Ruiz [1,2], Herbert Höpfl [2,*],
Hugo Morales-Rojas [2,*], Obdulia Sánchez-Guadarrama [2], Patricia Rodríguez-Cuamatzi [3]
and Dea Herrera-Ruiz [1,*]

[1] Facultad de Farmacia, Universidad Autónoma del Estado de Morelos, Av. Universidad 1001,
Cuernavaca 62209, Mexico
[2] Centro de Investigaciones Químicas, Instituto de Investigación en Ciencias Básicas y Aplicadas, Universidad
Autónoma del Estado de Morelos, Av. Universidad 1001, Cuernavaca 62209, Mexico
[3] Departamento de Ingeniería Química, Universidad Politécnica de Tlaxcala, Carretera Federal
Tlaxcala-Puebla Km 9.5, Tepeyanco, Tlaxcala 90180, Mexico
* Correspondence: hhopfl@uaem.mx (H.H.); hugom@uaem.mx (H.M.-R.); dherrera@uaem.mx (D.H.-R.)

Received: 22 November 2019; Accepted: 22 December 2019; Published: 25 December 2019

Abstract: The effect of hydroxypropyl methylcellulose (HPMC) and methylcellulose (Methocel® 60 HG) on the dissolution behavior of two cocrystals derived from nitazoxanide (NTZ), viz., nitazoxanide-glutaric acid (NTZ-GLU, 1:1) and nitazoxanide-succinic acid (NTZ-SUC, 2:1), was explored. Powder dissolution experiments under non-sink conditions showed similar dissolution profiles for the cocrystals and pure NTZ. However, pre-dissolved cellulosic polymer in the phosphate dissolution medium (pH 7.5) modified the dissolution profile of NTZ when starting from the cocrystals, achieving transient drug supersaturation. Subsequent dissolution studies under sink conditions of polymer-based pharmaceutical powder formulations with NTZ-SUC cocrystals gave a significant improvement of the apparent solubility of NTZ when compared with analogous formulations of pure NTZ and the physical mixture of NTZ and SUC. Scanning electron microscopy and powder X-ray diffraction analysis of samples recovered after the powder dissolution studies showed that the cocrystals undergo fast dissolution, drug supersaturation and precipitation both in the absence and presence of polymer, suggesting that the solubilization enhancement is due to polymer-induced delay of nucleation and crystal growth of the less soluble NTZ form. The study demonstrates that the incorporation of an appropriate excipient in adequate concentration can be a key factor for inducing and maintaining the solubilization of poorly soluble drugs starting from co-crystallized solid forms. In such a way, cocrystals can be suitable for the development of solid dosage forms with improved bioavailability and efficacy in the treatment of important parasitic and viral diseases, among others.

Keywords: nitazoxanide; cocrystals; multicomponent crystals; dissolution behavior; supersaturated formulations; crystallization inhibitors; drug-polymer interactions

---

1. Introduction

Nitazoxanide (NTZ), 2-acetyloxy-*N*-(5-nitro-2-thiazolyl)benzamide, is a synthetic nitrothiazole derivative (Scheme 1) with a broad spectrum of applications as an antiparasitic, antibacterial and antiviral agent, being effective against protozoal infections, helminths, gram negative and gram positive bacteria, and diverse viruses (respiratory viruses, rotavirus, norovirus, coronavirus, hepatitis B and C, dengue-2, yellow fever, Japanese encephalitis, and human immunodeficiency viruses) [1–3]. Moreover, NTZ has shown anticancer activity [4], suppresses the production of interleukins (IL) such as IL-6 [5], and is a promising compound for the treatment of neuropathic pain and the Ebola virus disease [6,7]. However, due to poor aqueous solubility (0.0075 mg/mL), NTZ has low bioavailability and requires high doses for treatment [8]. Considering its high permeability across intestinal epithelium [9], NTZ is a class II drug according to the criteria established by the Biopharmaceutical Classification System (BCS). To take advantage of the broad therapeutic spectrum of NTZ, the implementation of strategies to modify its solubility are necessary. A general strategy to overcome the limited solubility of drugs is the generation of novel solid phases, such as metastable amorphous, polymorphs, salts and, more recently, cocrystals [10–12].

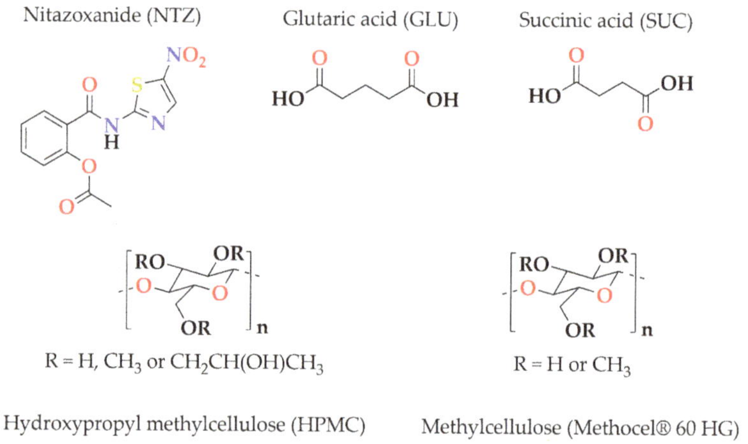

Scheme 1. Molecular structures of nitazoxanide (NTZ), coformers and cellulosic polymers used herein to achieve a polymer-based pharmaceutical powder formulation.

A cocrystal constitutes a solid single-phase crystalline material composed of two or more molecular and/or ionic compounds in a stoichiometric ratio, which is neither a solvate nor a simple salt. If at least one of the components is an active pharmaceutical ingredient (API) and if the coformer is pharmaceutically acceptable, then the substance is recognized as pharmaceutical cocrystal. A pharmaceutical cocrystal can offer multiple benefits for physicochemical and biopharmaceutical properties of APIs, such as increased solubility and dissolution rates, improvement of mechanical properties, and stability [13–15]. Particularly, cocrystals have exhibited higher, similar or lower solubility compared to the parent API, depending on the nature of the coformer and the solution conditions [16,17]. Solubility advantages of cocrystals over parent drugs can be higher by orders of magnitude, however, this also represents a risk due to precipitation to the less soluble form of the drug at supersaturated concentrations reached during the dissolution process [18,19].

Due to this former behavior, cocrystals are potential candidates for the development of supersaturating drug delivery systems [20,21]. A strategy to control cocrystal dissolution, drug supersaturation and precipitation is the use of supersaturated formulations, which incorporates additives and excipients such as cyclodextrins, surfactants and polymers [22]. In such systems, the supersaturation state must be maintained over a reasonable time period in order to promote adequate absorption for increased bioavailability. In this line, there are several recent reports on the effect of polymers and surfactants in increasing solubility and dissolution rates of poorly soluble drugs, starting from a cocrystalline solid phase. Some representative APIs and nutraceutical compounds studied under this approach are celecoxib [23], danazol [24], indomethacin [25], carbamazepine [26–31], flufenamic acid [32], cilostazol [33], resveratrol [34], dihydromyricetin [35], tadalafil [36], exemestane [37] and posaconazole [38].

Dissolution methods under non-sink conditions are commonly used for evaluating the ability of poorly soluble APIs to generate and to maintain a supersaturation state in solution [24,39]. Additionally, powder dissolution experiments are a good alternative to intrinsic dissolution tests for cocrystals undergoing rapid transformation to a less soluble phase of the drug [24], as recently found in the biopharmaceutical characterization of cocrystals with meloxicam, indomethacin, apixaban and myricetin [40–43].

Recently, the synthesis and characterization of NTZ cocrystals using succinic, glutaric, 2,5-dihydroxybenzoic, *p*-aminobenzoic and *p*-aminosalicylic acid as coformers were reported [8,44]. In both publications, improved intrinsic dissolution rates were documented, showing that NTZ cocrystals can improve the solubilization of NTZ in physiologically relevant media [8], thus being good candidates for the future development of novel dosage forms. Herein, two cocrystals, viz., NTZ-GLU and NTZ-SUC (GLU = glutaric acid; SUC = succinic acid), were chosen to evaluate the effect of two cellulosic polymers, namely hydroxypropyl methylcellulose (HPMC) and methylcellulose (Methocel® 60 HG) (Scheme 1), on the dissolution performance of the API, in comparison with pure NTZ, a commercially available NTZ medicine and the respective physical mixtures of NTZ and the coformer. Our study includes an API widely used in treatments concerning tropical neglected diseases, even though scarcely studied [45]. The purpose of this work was to study the behavior of NTZ cocrystals in solution and the conditions to sustain supersaturation levels since these conditions cannot be apparent and must be studied case-by-case. Given its therapeutic relevance, approaches to formulate nitazoxanide cocrystals will be of benefit to improve its efficacy, considering that its low solubility impacts its bioavailability. Importantly, nitazoxanide has been proposed as a potential agent for public health control [46].

2. Materials and Methods

2.1. Materials

Nitazoxanide was kindly donated by Laboratories Senosiain, S.A de C.V., CDMX, México. Glutaric acid, succinic acid, hydroxypropyl methylcellulose (HPMC, MW: 26 kDa. This product has a methoxyl content of 19–24% and a hydroxypropoxyl content of 7–12%, viscosity 80–120 cP, 2% in H_2O at 20 °C), methylcellulose (Methocel® 60 HG and Methocel® MC), hydroxypropyl cellulose (HPC 80,000 and HPC 370,000), polyvinylpyrrolidone (Kollidon® 25), polyethylene glycol (Kollisolv®) and poly(ethylene glycol)-block-poly(propylene glycol)-block-poly(ethylene glycol) (Kolliphor® P 407) were purchased from Sigma-Aldrich (St. Louis, MO, USA) as reagents for use only in the laboratory. All organic solvents used were purchased from J. T. Baker Chemical Co. (Phillipsburg, NJ, USA). Solvents and chemical compounds were used as received without further purification.

2.2. Methods

2.2.1. Preparation of NTZ Cocrystals

NTZ-GLU and NTZ-SUC cocrystals of 1:1 and 2:1 NTZ:coformer stoichiometric ratio, respectively, were synthetized by the liquid-assisted grinding (LAG) technique as previously reported [8]. Samples of 1.0 g were ground in a Retsch MM400 mixer (Haan, Germany; 0.5 h at 25 Hz) with acetone. The formation of the cocrystalline solid phases was established by comparison of the powder X-ray diffraction (PXRD) patterns with both experimental and simulated patterns obtained from reported X-ray crystal structures (REFCODES: BREEINM and BRENVIEX).

2.2.2. Solvent-Shift Method

A total of eight candidate polymers were evaluated to test their ability to maintain NTZ solubilized: Methocel® MC, Methocel® 60 HG, HPMC, HPC 80,000, HPC 370,000, Kollisolv®, Kolliphor® P 407, and Kollidon® 25 (see Table S1, for further polymer details). The tests were performed by the modified solvent shift method proposed by Childs et al. in 2013 [24], see Figure S1. Aqueous solutions were prepared containing 0.5% *w/v* of each pre-dissolved polymer in pH 7.5 phosphate buffer solution (PBS) and also one containing only buffer solution. A volume of 3 mL of each medium was placed in a spectrophotometer quartz cell, followed by a stepwise addition of a total of 13 aliquots of 10 µL each from NTZ stock solution in DMSO (25 mg/mL) at 5 min interval times. The ability of the polymeric compounds to maintain the API dissolved was determined by analysis of the presence or absence of precipitated NTZ in solution, which can be monitored by UV-vis spectrophotometry measured at $\lambda = 550$ nm.

2.2.3. Powder Dissolution under Non-Sink Conditions

For the powder dissolution experiments under non-sink conditions, 200 mg of pure NTZ or the equivalent of NTZ cocrystal were added to 10 mL of pH 7.5 buffer dissolution medium. All NTZ samples were passed through a sieve mesh 200.

The dissolution experiments were carried out at 37 ± 0.5 °C with magnetic stirring at 90 rpm (Personal Reaction Station, J-Kem Scientific Inc., St. Louis, MO, USA). Samples of 1 mL were withdrawn and subsequently filtered (Whatman 3), each minute during the first 10 min, then every two minutes until 30 min, and finally at 45, 60, 75 and 90 min. A total of 200 µL samples were then diluted to 5 mL with the corresponding dissolution medium and analyzed by UV-vis spectrophotometry at 435 nm to determine the amount of NTZ dissolved. Dissolution test with solutions containing pre-dissolved HPMC or Methocel® 60 HG polymer (0.5% *w/v*) in pH 7.5 PBS were employed and compared with the results obtained from polymer-free solutions (PBS alone). All experiments were carried out in triplicate. After the dissolution experiments, the solid residues were collected and dried at room temperature for analysis by powder X-ray diffraction (PXRD).

2.2.4. Preparation of Powder Formulations with Methocel® 60 HG

Powder formulations with NTZ consisted in physical mixtures of the components listed in Table 1. Formulations at three different concentrations (1.0, 2.5 and 5.0% *w/w*) of Methocel® 60 HG were prepared considering 250 mg of nitazoxanide in the powder mix (entries 4 to 12). For comparison, the corresponding control experiments in the absence of Methocel® 60 HG were also obtained (entries 1, 2 and 3). All NTZ samples were passed through a sieve mesh 200.

Table 1. Composition (mg) of powder formulations with and without Methocel® 60 HG for NTZ, the physical mixture of NTZ and succinic acid (SUC) (2:1), and the NTZ-SUC cocrystal (2:1) [a].

Formulation	1	2	3	4	5	6	7	8	9	10	11	12
NTZ	250			250			250			250		
Phys. Mix. NTZ/SUC		298			298			298			298	
NTZ-SUC cocrystal			298			298			298			298
Methocel® 60 HG [b]	0	0	0	3.8	3.8	3.8	9.5	9.5	9.5	19	19	19

[a] 298 mg of NTZ-SUC (2:1) is the molar equivalent of 250 mg of NTZ. [b] Percentage (*w/w*) of the polymer in the formulation was calculated by taking as reference the average weight of the commercial tablet (ca. 758 mg) containing 500 mg of NTZ. The equivalent weight of the tablet for 250 mg of NTZ corresponds to 379 mg.

2.2.5. Powder Dissolution Experiments in the USP 1 (Basket) Apparatus

For the dissolution experiments employing the USP 1 apparatus, the powder formulations shown in Table 1 were added to a volume of 600 mL of pH 7.5 PBS solution, with a rotation speed of 100 rpm at 37 ± 0.5 °C. Samples of 4 mL were taken at 2, 5, 10, 15, 30, 60, 120 and 180 min, and an equal volume of fresh medium was added to maintain the dissolution medium volume constant. The NTZ concentration was quantified by UV-vis spectrophotometry at 435 nm (Figures S2 and S3). Each dissolution profile represents the average of three experiments. After the dissolution experiments, the solid residues were collected and dried at room temperature for analysis by PXRD and by scanning electron microscopy (SEM). Finally, a test comparing a powder formulation with Methocel® 60 HG 5% *w/w* containing 500 mg of NTZ as NTZ-SUC cocrystal and a commercially available formulation

(500 mg tablet) was performed. The reference tablets of NTZ (Daxon®, 500 mg NTZ) were ground in an agate mortar and passed through a sieve mesh 200. All the dissolution profiles are presented as mg of NTZ dissolved vs. time. The mg of NTZ dissolved was calculated as:

$$\text{mg of drug dissolved} = C_n \times V_n + \sum_{i=1}^{n-1} C_i \times V_s$$

where C_n = concentration of drug in sample n, V_n = volume of dissolution media at the time of taking the sample n; C_i = concentration of drug in sample $n-1$ and V_s = volume of aliquot due to sampling [47].

2.3. Characterization Techniques

2.3.1. Powder X-Ray Diffraction Analysis (PXRD)

The solid phases resulting from the NTZ cocrystal synthesis and the samples recovered from the powder dissolution experiments were examined by PXRD analysis using a BRUKER D8-ADVANCE diffractometer (Bruker, Bremen, Germany) equipped with a LynxEye detector ($\lambda_{CuK\alpha1}$ = 1.5406 Å, monochromator: germanium). The equipment was operated at 40 kV and 40 mA, and data were collected at room temperature in the range of 2θ = 5–45°. Powder diffraction patterns were compared with the simulated patterns from single crystal X-ray diffraction (SCXRD) data.

2.3.2. Scanning Electron Microscopy (SEM) Analysis

The solid residuals obtained after the dissolution tests under sink conditions were analyzed by SEM. For this purpose, the samples were coated with a gold layer, where after SEM micrographs were captured by a VEGA 3 TESCAN scanning electron microscope (Tescan Orsay Holdings, Kohoutovice, Czech Republic) at 10.0 kV.

2.4. Statistical Analysis

The differences in the dissolution profiles were analyzed by two-way variance analysis (ANOVA) of the area under the curve (AUC), using the Origin Pro 9.0 software package (OriginLab Co., Northampton, MA, USA). Subsequently, a multiple post hoc test (Tukey) was performed with a significance level of 0.05 using the Statgraphics software (Statgraphics Technologies, Inc., The Plains, VA, USA).

3. Results

3.1. Preparation and Characterization of NTZ Cocrystals

NTZ was combined with GLU and SUC in 1:1 and 2:1 stoichiometric ratio, respectively, using the liquid-assisted grinding method with acetone as solvent as previously reported [8]. NTZ-SUC and NTZ-GLU cocrystals were prepared in scales of 1.0 g, and in all cases a homogeneous single phase was obtained, for which the experimental PXRD pattern agreed with the pattern simulated from the crystal structure determined by single-crystal X-ray diffraction analysis (see Figure S4). A batch of ten (individually characterized) samples was finally mixed and used for all subsequent experiments.

3.2. Polymer Selection by Solvent-Shift Method

Figure 1 shows the effect of polymers commonly employed in pharmaceutical drug formulations to inhibiting or delaying precipitation of NTZ. In the absence of polymers, addition of aliquots from NTZ stock solution (25 mg/mL) to 3 mL of a phosphate buffer solution (PBS, pH 7.5) caused precipitation of NTZ at concentrations above 0.30 mg/mL. In the presence of 0.5% *w/v* pre-dissolved cellulosic polymers, i.e., Methocel® MC, Methocel® 60 HG, HMPC, HPC 80,000 and HPC 370,000, as well as the polyvinylpyrrolidone Kollidon® 25, precipitation of NTZ was inhibited, reaching concentrations of at least 1.0 mg/mL; meanwhile, polyethylene glycol Kollisolv® and poloxamer Kolliphor® P 407 maintained the drug dissolved only at concentrations below 0.25–0.40 mg/mL. Based on these results, HPMC and Methocel® 60 HG were selected for subsequent powder dissolution experiments of NTZ and the cocrystal phases with GLU and SUC under non-sink conditions.

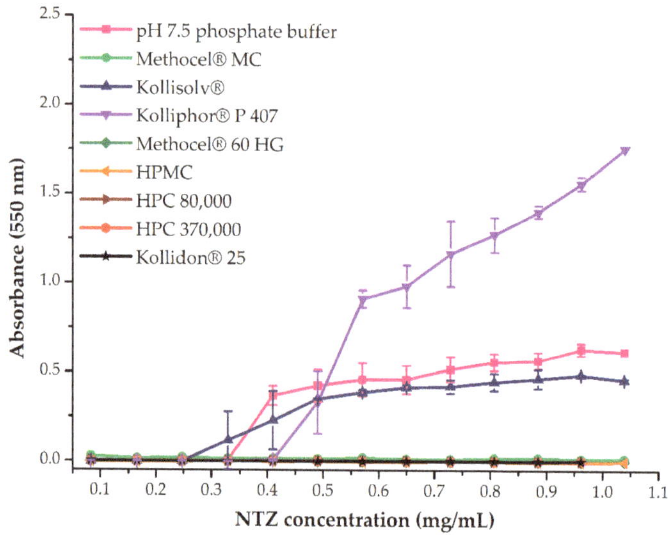

Figure 1. Inhibitory effect of pre-dissolved polymers (0.5% *w/v*) on NTZ precipitation in pH 7.5 PBS. (*n* = 3 ± SD).

3.3. Powder Dissolution under Non-Sink Conditions

Powder dissolution profiles of pure NTZ and the NTZ-GLU/NTZ-SUC cocrystals were measured in pH 7.5 phosphate buffer solution in the absence and presence of pre-dissolved polymer (0.5%, *w/v*; 5 mg/mL). The corresponding graphs are shown in Figure 2. In the absence of polymer, there is no statistically significant difference ($p > 0.05$) between the profile of NTZ and those obtained for the cocrystals (Figure 2a), maintaining on average a saturated NTZ concentration of 0.5 mg/mL. On the contrary, in the presence of pre-dissolved HPMC and Methocel® 60 HG, the NTZ dissolution profiles of the pure form and the cocrystals are significantly different ($p < 0.05$) (Figure 2b,c). Thus, starting from the cocrystals, pre-dissolved HPMC and Methocel® 60 HG enable the generation of transient drug concentrations (approximately 2 mg/mL) that are 4 times above NTZ solubility, following a spring–parachute profile as postulated by Guzman et al. [48]. This supersaturation state is sustained for approximately 30 min, for both cocrystals. The decay of the NTZ concentration is attributed to precipitation of pure NTZ, as indicated by PXRD analyses of the solid residues recovered after the dissolution experiments with NTZ-GLU and NTZ-SUC, that showed characteristic diffraction peaks of pure NTZ (Figures 3–5). Overall, this analysis shows that indeed NTZ-GLU and NTZ-SUC have superior dissolution properties compared to the parent drug, suggesting that dissolution kinetics of the cocrystalline phases is extremely rapid, supersaturating the solution and precipitating in a short time period (~1–5 min), which is in agreement with our previous studies on the solution-phase stability of NTZ-GLU and NTZ-SUC in PBS pH 7.5 [8]. However, in the presence of pre-dissolved HPMC and Methocel® 60 HG, the precipitation of pure NTZ is delayed by a time interval long enough to envision a potential improvement of the bioavailability (~30 min, see Discussion section).

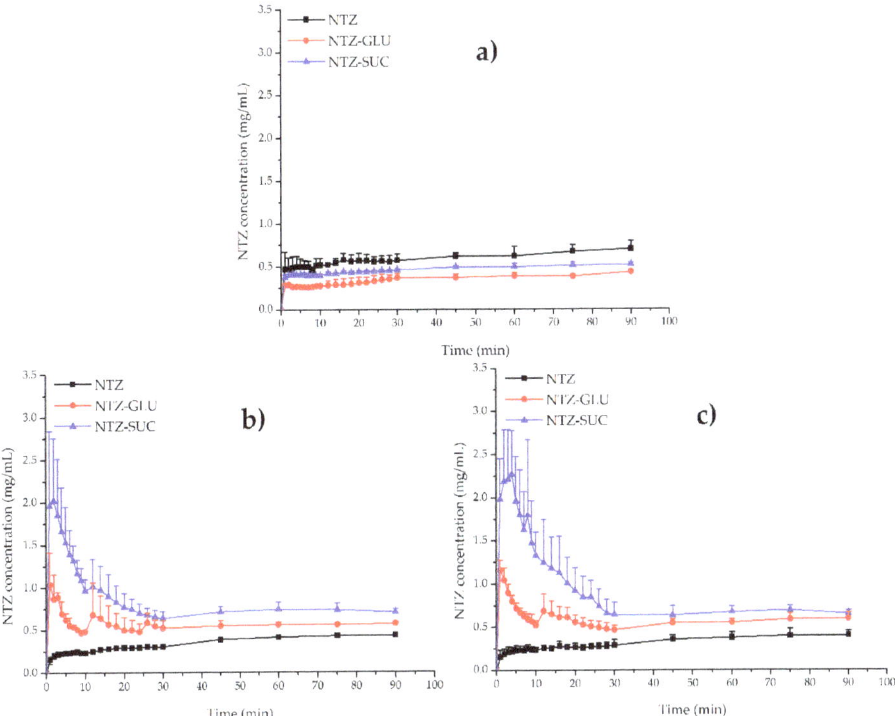

Figure 2. Powder dissolution profiles of NTZ and the cocrystalline phases NTZ-GLU and NTZ-SUC in: (**a**) pH 7.5 PBS; (**b**) PBS with pre-dissolved HMPC (0.5%, *w/v*); and (**c**) PBS with pre-dissolved Methocel® 60 HG (0.5%, *w/v*) (*n* = 3 ± SD).

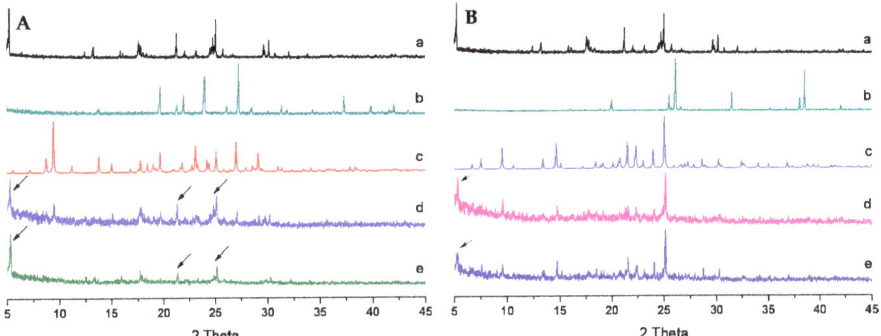

Figure 3. Comparison of powder X-ray diffraction (PXRD) patterns: (**A**) (a) NTZ, (b) GLU, (c) NTZ-GLU simulated from SCXRD data. Solids recovered after powder dissolution tests of NTZ-GLU under non-sink conditions in pH 7.5 phosphate buffer solution after (d) 2 min, and, (e) 5 min. (**B**) (a) NTZ, (b) SUC, (c) NTZ-SUC simulated from SCXRD data. Solids recovered after powder dissolution tests of NTZ-SUC under supersaturated conditions in pH 7.5 phosphate buffer solution after (d) 1 min, and (e) 3 min.

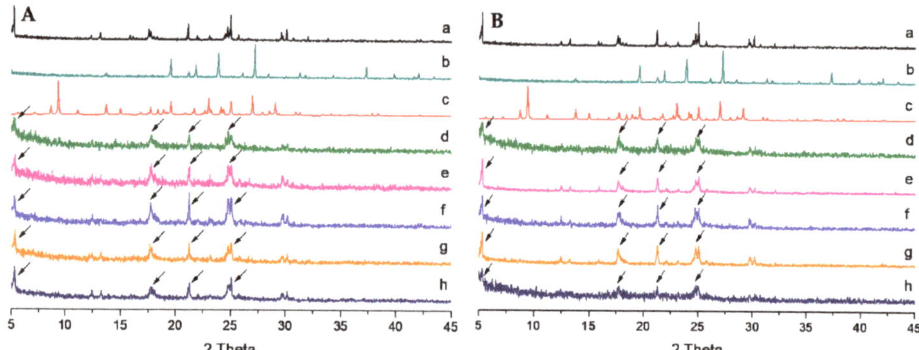

Figure 4. Comparison of PXRD patterns: (**A**) (a) NTZ, (b) GLU, (c) NTZ-GLU simulated from SCXRD data. Solids recovered after powder dissolution tests of NTZ-GLU under non-sink conditions in pH 7.5 phosphate buffer solution with pre-dissolved 0.5% *w/v* HPMC after (d) 1 min, (e) 2 min, (f) 3 min, (g) 4 min, (h) 5 min (**B**) (a) NTZ, (b) GLU, (c) NTZ-GLU simulated from SCXRD data. Solids recovered after powder dissolution tests of NTZ-GLU under non-sink conditions in pH 7.5 phosphate buffer solution with pre-dissolved 0.5% *w/v* Methocel® 60 HG after (d) 1 min, (e) 2 min, (f) 3 min, (g) 4 min, (h) 5 min.

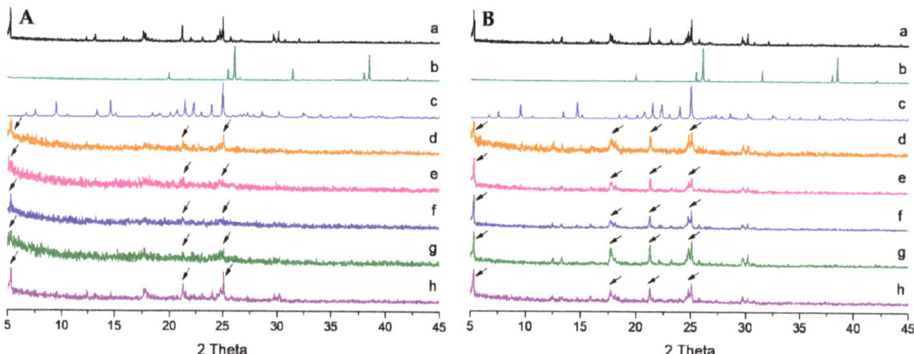

Figure 5. Comparison of PXRD patterns: (**A**) (a) NTZ, (b) SUC, (c) NTZ-SUC simulated from SCXRD data. Solids recovered after powder dissolution tests of NTZ-SUC under non-sink conditions in pH 7.5 phosphate buffer solution with pre-dissolved 0.5% *w/v* HPMC after (d) 1 min, (e) 2 min, (f) 3 min, (g) 4 min, (h) 5 min. (**B**) (a) NTZ, (b) SUC, (c) NTZ-SUC simulated from SCXRD data. Solids recovered after powder dissolution tests of NTZ-SUC under non-sink conditions in pH 7.5 phosphate buffer solution with pre-dissolved 0.5% *w/v* Methocel® 60 HG after (d) 1 min, (e) 2 min, (f) 3 min, (g) 4 min, (h) 5 min.

3.4. USP 1 Apparatus Powder Dissolution Experiments of Formulations with Polymer

The above-described powder dissolution experiments under non-sink conditions showed that the best improvement of the NTZ dissolution profile was achieved with the NTZ-SUC cocrystal, in the presence of Methocel® 60 HG. To evaluate if the polymer included in the solid formulation has an effect on the dissolution kinetics of the cocrystal, four pharmaceutical powder formulations were examined with cocrystal–polymer ratios of 0.0, 1.0, 2.5 and 5.0% (*w/w*). The dissolution performance of the formulated solid samples was analyzed by means of powder dissolution tests using the USP 1 (basket) apparatus. A similar strategy was employed for the evaluation of formulated cocrystals of carbamazepine [29,30]. The dissolution profiles in pH 7.5 PBS for NTZ in pure form, a 2:1 stoichiometric physical mixture of NTZ and SUC, and NTZ-SUC (2:1) cocrystals formulated with an increasing amount of Methocel® 60 HG are given in Figure 6.

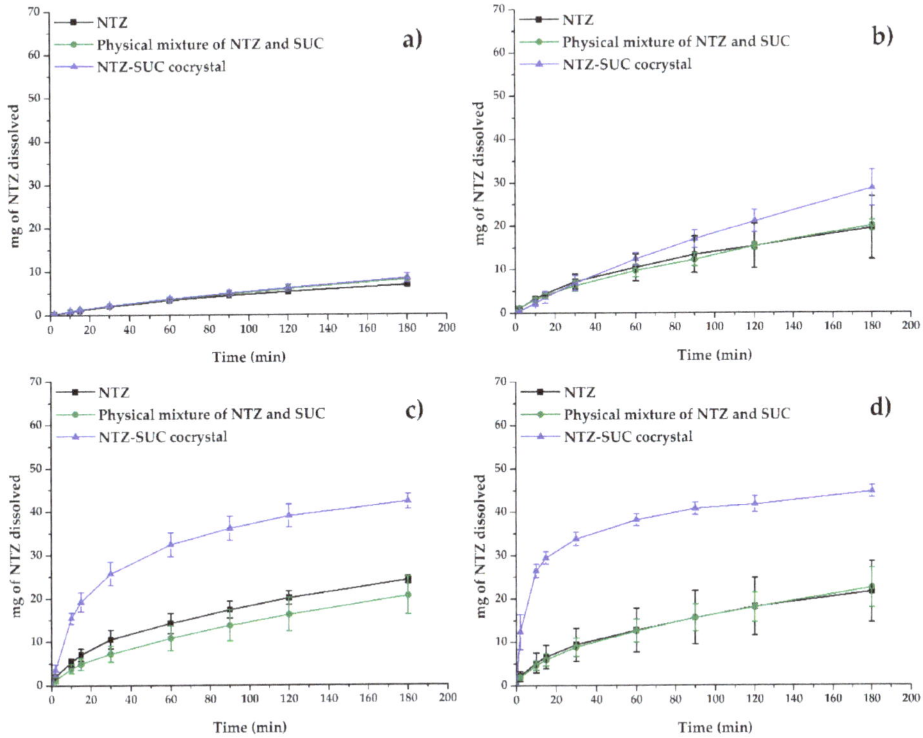

Figure 6. Powder dissolution profiles (in pH 7.5 PBS at 37 ± 0.5 °C) for pure NTZ, the 2:1 stoichiometric physical mixture of NTZ and SUC, and NTZ-SUC (2:1) cocrystal formulated with Methocel® 60 HG (*w/w*): (**a**) 0.0%; (**b**) 1.0%; (**c**) 2.5%; (**d**) 5.0% (*n* = 3 ± SD).

Figure 6a illustrates the dissolution profiles measured over a time interval of 3 h of the unformulated (0.0% *w/w* of Methocel® 60 HG) solid phases of pure NTZ, the physical mixture with SUC and the NTZ-SUC cocrystal. The drug dissolution profiles are similar ($p > 0.05$), achieving approximately 8 mg of dissolved drug (ca. 3% in relation to 250 mg of NTZ dose). In contrast, when the solid phases were formulated with 1.0% Methocel® 60 HG, the drug was dissolved in significantly larger amounts, either starting from the pure form, the physical mixture or the cocrystal (~17–27 mg of drug dissolved after 3 h). Although the graph associated to NTZ-SUC exhibits a larger increase, the statistical difference ($p > 0.05$) between the dissociation rates is not significant due to the relatively large standard deviation (Figure 6b). However, with larger amounts of Methocel® 60 HG polymer (2.5 and 5.0%), the NTZ-SUC cocrystal gave a significantly improved dissolution profile ($p < 0.05$) when compared to analogous formulations of pure NTZ and the physical mixture, achieving after 3 h, amounts of 42 and 45 mg of NTZ dissolved (Figure 6c,d). Furthermore, for the formulation of the cocrystal with 5% Methocel® 60 HG, the amount of NTZ dissolved after 2 min was already approximately 12 mg (Figure 6d), which is significantly larger than the final amount achieved by the corresponding unformulated solid phases (Figure 6a).

The cocrystal solubilization can be quantified by the AUC (area under the curve), thus constituting a parameter indicative of the influence that cocrystals have on the API performance. The AUC is directly and inversely proportional to the dissolution and precipitation rates, respectively [49]. Figure 7 shows that unformulated NTZ-SUC cocrystals and NTZ-SUC formulated with 1.0% *w/w* Methocel® 60 HG do not lead to improved dissolution profiles when compared to NTZ and the physical mixture of NTZ and SUC ($p > 0.05$). However, for the formulations with 2.5 and 5.0% (*w/w*) of polymer, the AUC for the NTZ-SUC cocrystal increased by factors of 8.0 and 9.2, respectively, in comparison to the unformulated NTZ. Meanwhile, for NTZ and the physical mixture with SUC the AUC are practically constant in the polymer formulations examined herein.

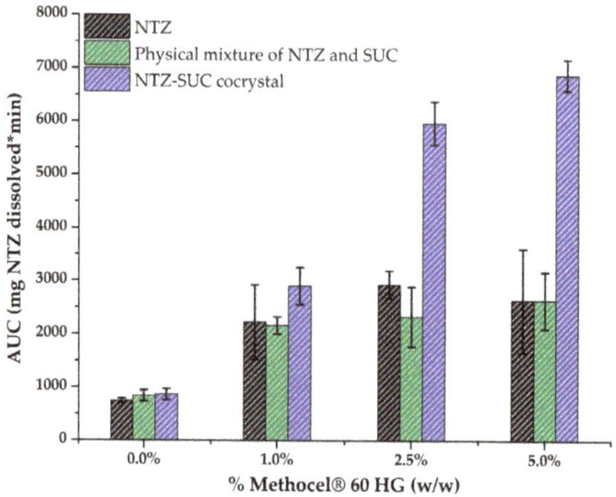

Figure 7. AUC of NTZ extracted from the dissolution profiles in pH 7.5 PBS for the following solid phases, which were formulated with different amounts of Methocel® 60 HG (0.0, 1.0, 2.5 and 5.0%, *w/w*): NTZ, 2:1 physical mixture of NTZ and SUC, and NTZ-SUC cocrystal.

Selected samples of the solid residues recovered after the dissolution tests were analyzed by PXRD and SEM analysis. The PXRD data showed that the NTZ-SUC cocrystal is transformed into its parent drug in all of the formulations tested (Figure 8), and it was also observed in the dissolution experiments under non-sink conditions. Peaks for SUC are not observed, indicating that the coformer was completely dissolved, which is in agreement with its good solubility in water (71 mg/mL at 25 °C).

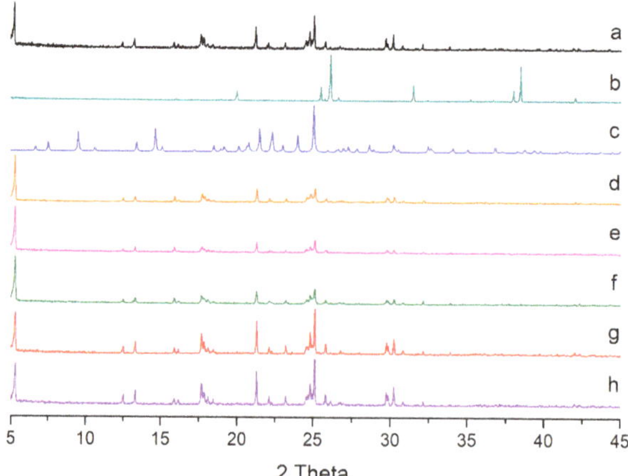

Figure 8. PXRD patterns of (a) NTZ, (b) SUC, (c) NTZ-SUC cocrystal (simulated from SCXRD data) and solid residues after dissolution tests in USP 1 apparatus: (d) NTZ-SUC formulated with 1.0% *w/w* Methocel® 60 HG, (e) NTZ-SUC formulated with 2.5% *w/w* Methocel® 60 HG, (f) NTZ-SUC formulated with 5.0% *w/w* Methocel® 60 HG, (g) NTZ formulated with 5.0% *w/w* Methocel® 60 HG, and (h) physical mixture of NTZ and SUC formulated with 5.0% *w/w* Methocel® 60 HG.

Figure 9 presents SEM images of the cocrystal phase NTZ-SUC before and after the dissolution experiments in the presence of different amounts of Methocel® 60 HG. Freshly prepared NTZ-SUC cocrystals exhibit a characteristic and homogeneous cylindrical morphology with a small particle size (Figure 9a) in comparison with pure NTZ and the physical mixture with SUC (Figure S5a,b). The SEM images of the solid residues recovered after the dissolution tests of NTZ and the physical mixture formulated with Methocel® 60 HG at 1.0 and 5.0% (*w/w*) show slight changes of the morphology in the presence of polymer (Figure S5c–f). A more pronounced effect occurs for the NTZ-SUC cocrystals with the morphology of the samples changing significantly with increasing amount of polymer (Figure 9b–d).

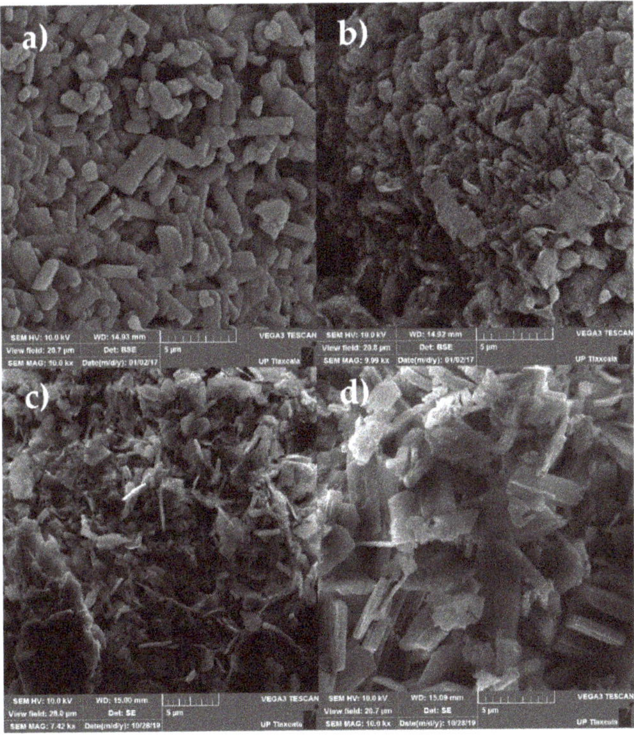

Figure 9. SEM images of (**a**) NTZ-SUC cocrystals before the powder dissolution test, and the solid residues recovered after the dissolution experiments in the USP 1 apparatus: NTZ-SUC cocrystal formulated with (**b**) 1.0%, (**c**) 2.5%, and (**d**) 5% of Methocel® 60 HG.

Because of the promising results, the performance of the powder formulation of NTZ-SUC cocrystal with Methocel® 60 HG (5% *w/w*) was compared with a commercially available formulation of NTZ. Since our studies were based on a pharmaceutical powder, the commercially available reference tablets of NTZ (Daxon®, 500 mg NTZ) were ground in an agate mortar and sifted through a sieve mesh 200. The results of the powder dissolution experiments using the USP type 1 apparatus (following the protocol detailed in Section 2.2.5) are shown in Figure 10. Comparison of the dissolution graphs shows a statistically significant improvement of the dissolution properties for the formulated NTZ-SUC cocrystal in comparison with the commercial NTZ medicine ($p < 0.05$). From the polymer–cocrystal formulation with Methocel® 60 HG (5% *w/w*), 57 mg of the drug were dissolved in the bulk solution after 3 h with an amount dissolved of 9046 ± 533 mg NTZ*min ($AUC_{0-3\,h}$), while the commercial product achieved 50 mg in the same time period, with an amount of 7287 ± 77 mg NTZ*min dissolved.

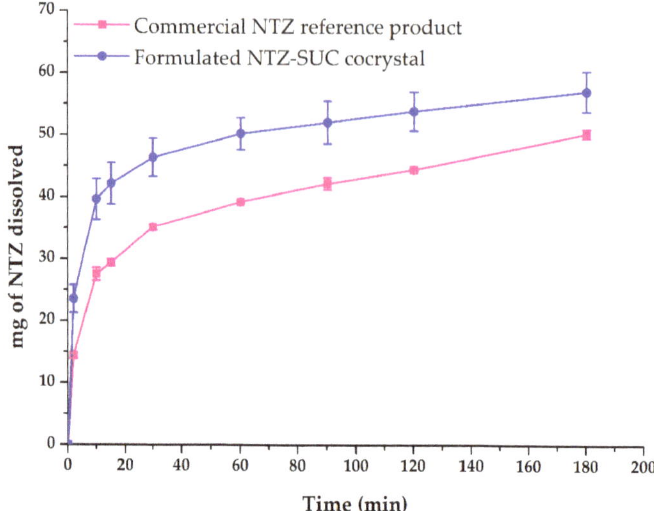

Figure 10. Powder dissolution profiles of a commercially available formulated NTZ medicine and the formulated powder of NTZ-SUC cocrystals with Methocel® 60 HG (5% *w/w*) in pH 7.5 PBS at 37 ± 0.5 °C (*n* = 3 ± SD).

4. Discussion

In a recent review Zaworotko and coworkers enlisted a total of eight commercially available pharmaceutical cocrystals [50], showing that cocrystalline solid forms of current APIs are adequate for the fabrication of medicines for various types of therapies. At the same time, in the last few years the number of publications and patents related to pharmaceutical cocrystals has been increasing [50,51]. Nevertheless, there are still several challenges for the development of cocrystals in pharmaceutical solid dosage forms, particularly in controlling the behavior of cocrystal dissolution, drug supersaturation and precipitation kinetics, which can be achieved by selecting suitable excipients that will delay nucleation and/or crystallization, and maintain adequate drug supersaturation levels [13,51].

In the research project presented herein, two cellulosic polymers were chosen to evaluate their effect on maintaining the supersaturated state achieved during the dissolution of cocrystalline solid phases of nitazoxanide, a poorly soluble drug. Subsequently, polymer–cocrystal powder formulations were developed. To select adequate candidates from the series of widely used polymers in the pharmaceutical industry that delay the nucleation and/or crystallization of NTZ in supersaturated conditions, the solvent-shift approach was used [24], see Figure S1. Figure 1 demonstrates the results with the polymers tested for NTZ in pH 7.5 PBS at room temperature, finding that non-cellulosic polymers (Kollisolv® and Kolliphor® P407) do not significantly help in delaying NTZ precipitation, while polyvinylpyrrolidone (PVP-Kollidon® 25) and cellulosic polymers (HPC 80,000, HPC 370,000, HPMC, Methocel® MC and Methocel® 60 HG) enable the increase of the concentration of NTZ in aqueous solution up to at least 1.0 mg/mL. Considering the hydrogen bonding capacity, NTZ has one H-bond donor group (NH) and four H-bond acceptor sites [8]. As can be observed in Scheme 1, the cellulosic polymers HPMC and Methocel® 60 HG contain a large number of O–H donor groups, which are complementary for the formation of hydrogen bonds with hydrogen bond acceptors, explaining the good precipitation inhibitor properties for NTZ. The remaining polymers tested herein lack strong H-bond donor functions that could interact through hydrogen bonds with NTZ. These findings are supported by previous reports in the literature, e.g., in 2010, Warren et al. described the effectiveness of cellulosic polymers as crystallization inhibitors for danazol [52]. Other reports showed that HPMC reduces efficiently the nucleation and crystal growth of felodipine [53] and celecoxib [54] under non-sink conditions. Moreover, it is reasonable to assume that cellulose derivatives do inhibit, not only nucleation but also crystal growth and agglomeration into larger sized particles (see SEM results) that otherwise may form from during NTZ precipitation. In other words, the precipitation delay of NTZ can also be attributed to surface interactions between the NTZ particles and the polymer, forming drug-polymer complexes, which is derived in the modification of the NTZ solubilization properties.

4.1. Performance of NTZ Cocrystals under Non-Sink Conditions

Based on the large solubility difference between NTZ and aliphatic dicarboxylic acids as coformers [8], we expected that cocrystal forms of NTZ with GLU and SUC increase the aqueous solubility of NTZ. Figure 2a shows that NTZ-GLU and NTZ-SUC cocrystals have dissolution profiles similar to NTZ in pH 7.5 phosphate buffer solution, and a higher solubilization of NTZ is not observed under these conditions. Indeed, rapid conversion of the cocrystals into pure drug was demonstrated by PXRD analysis of the solids recovered after the dissolution tests in short time periods (~1–5 min, Figure 3). A similar behavior has been documented for other cocrystal phases in the literature [24,25,30,36,37,55–57]. For example, exemestane–maleic-acid cocrystals did not increase solubilization of the drug due to rapid phase transformation [55]. In a recent publication this cocrystal was reinvestigated observing that transformation into the parent drug occurred in less than 1 min in fasted state simulated intestinal fluid (FaSSIF) [37]. The critical role of the coformer physicochemical properties on the solubility and dissolution advantages of cocrystals it is now accepted, even though the exact mechanism by which this enhancement occurs is not fully understood. In NTZ cocrystals studied here, the coformers GLU and SUC are ionizable compounds (pK_a values of 4.3 for GLU and 4.2 for SUC) and exhibit favorable water solubilities of 540 mg/mL and 71 mg/mL at 25 °C, respectively [12,58]. The fast dissolution–supersaturation phenomenon induced for the NTZ cocrystals under the experimental conditions used herein (pH 7.5 PBS), is attributed to the favorable solubilization of these coformers. This rapid cocrystal dissolution produces drug supersaturation levels that are difficult to control in absence of suitable excipients at adequate concentrations, and therefore NTZ solubilization advantage was not observed.

The use of excipients such as polymers and surfactants constitutes an efficient strategy to prolong drug supersaturation achieved from pharmaceutical cocrystals [23–25,36,57]. An addition of a cellulosic polymer into the dissolution medium for powder dissolution studies under non-sink conditions revealed that these can induce a supersaturation effect starting from NTZ-SUC and NTZ-GLU cocrystals. Both Methocel® 60 HG and HPMC were able to maintain drug supersaturation (up to 4 times higher concentrations of NTZ in solution over solubility value) for a period long enough to display a superior dissolution profile (Figure 2b,c). The phenomenon was upheld for up to 30 min, emulating the spring–parachute effect [48]. There are some studies demonstrating an in vitro/in vivo correlation of formulated cocrystals. In these cases, the in vitro dissolution test showed that the high drug concentration achieved by cocrystalline phases is maintained by 30–40 min and the bioavailability of the corresponding poorly soluble drugs was enhanced [24,33]. Considering these experimental data, it is considered that the increase of NTZ dissolved from NTZ cocrystals formulated with polymers may promote NTZ absorption.

The physical stability of the NTZ cocrystals used in the powder dissolution experiments was monitored by PXRD analysis, revealing that rapid cocrystal dissolution followed by drug precipitation still occurred in the presence of the polymer in solution (Figures 4 and 5). After 1 min in contact with pH 7.5 PBS, the powders analyzed by PXRD analysis indicated complete transformation to pure NTZ, and the patterns did not exhibit peaks indicating the presence of coformer in the solid.

Cellulosic polymers are the most common hydrophilic polymers used for the development of formulations with cocrystals [24,26,28,29,33,36,37]. They can inhibit or at least delay phase transformations of cocrystals [29,30], by establishing drug-polymer interactions [37] and maintaining the solubilization state of APIs. Therefore, the improvement of the NTZ dissolution profiles in the presence of pre-dissolved polymer could be explained by the following mechanism: after the rapid NTZ release into solution starting from the cocrystal, the cellulosic polymer interacts with NTZ molecules delaying its nucleation and eventually NTZ crystallized also interact with polymer interfering the crystal growth and agglomeration process.

4.2. NTZ-SUC Cocrystal Formulation vs. Formulation with Pure NTZ

Previous formulations with cocrystals were based on polymer matrix tablets [29,30] and pre-dissolved polymers [24,36]. However, there are few studies which have tested the effect of a polymer incorporated in the powder solid-state form on the dissolution profiles of cocrystals [23]. In order to evaluate the effect of Methocel® 60 HG on the dissolution process of NTZ-SUC cocrystals starting from a solid polymer–cocrystal formulation, we examined a series of pharmaceutical powder-based formulations containing different amounts of Methocel® 60 HG (0.0, 1.0, 2.5 and 5.0% *w/w*) in comparison to analogous forms with pure NTZ and the physical mixture of NTZ and SUC. In the absence of polymer, all solid phases released similar amounts of NTZ into the solution during the powder dissolution studies using the USP 1 apparatus (Figure 6a) and the NTZ-SUC cocrystal did not show a solubility advantage in comparison with the pure drug and the physical mixture. This observation can be attributed to immediate conversion of the cocrystal phase into NTZ and release of SUC into the solution, in accordance with the dissolution experiments under non-sink conditions. Formulations of pure NTZ, the physical mixture with SUC, and NTZ-SUC cocrystals with 1% (*w/w*) of Methocel® 60 HG showed dissolution profiles being similar among each other, but at higher concentrations, i.e., 2.5 and 5.0% (*w/w*), a significant increase of the dissolution rate was observed (Figure 6c,d and Figure 7).

Despite the inclusion of polymer, the NTZ-SUC cocrystals were transformed fast into pure NTZ, as revealed by PXRD patterns of solid residues recovered after the dissolution tests (Figure 8). However, the SEM analysis reveals that methylcellulose affects significantly the particle size and morphology of NTZ (Figure 9). The polymer–NTZ interactions probably influence the nucleation process and subsequent crystal growth/agglomeration, thus giving high drug concentration in the dissolution medium over a prolonged time period. Although NTZ-polymer interactions are expected to operate also for the formulations with pure NTZ and the physical mixture, these formulations lack the advantageous initial and immediate dissolution increase of the cocrystal components driven by favorable coformer solubilization.

Finally, comparison of the dissolution profiles of NTZ-SUC cocrystals formulated with methylcellulose at 5.0% *w/w* with a commercially available NTZ medicine showed that pharmaceutical products containing co-crystallized APIs are promissory. In such a way, reduction of the NTZ dose might become feasible, which would be a large advantage for the administration of this antiparasitic drug, which requires large doses. However, we are aware that to draw conclusions about this possibility and to assure therapeutic efficacy with reduced adverse effects, in vivo studies are required.

5. Conclusions

The results presented herein have shown that NTZ cocrystals can induce a superior dissolution behavior in the presence of hydroxypropyl methylcellulose (HMPC) and methylcellulose (Methocel® 60 HG) in comparison to pure NTZ. Although rapid transformation to the less soluble drug form is observed, the inclusion of polymer in the formulations with NTZ-SUC cocrystals in amounts of 2.5 and 5.0% (*w/w*) enabled significantly increasing the dissolution rate of NTZ when compared to pure NTZ and the commercially available drug formulation. The comparative analysis of different powder formulations showed that the amount of polymer in the formulation is an important parameter to consider during cocrystal formulation development for accessing a good solubilization potential for poorly soluble APIs in aqueous solutions.

In conclusion, the use of cocrystals combined with an appropriate selection of pharmaceutical excipients as drug precipitation inhibitors or retardants for crystal growth, such as cellulosic polymers, is a feasible approach to modify the dissolution environment and conditions starting from cocrystals.

As a result, the development of a polymer–cocrystal powder formulation is a true alternative to enhance the solubilization of poorly water soluble NTZ and to generate solid dosage forms with better bioavailability and efficacy in the treatment of parasitic and viral diseases, among others.

Supplementary Materials: The following data are available online at http://www.mdpi.com/1999-4923/12/1/23/s1: Table S1. General characteristics of polymers used. Figure S1. Scheme of the solvent-shift methodology employed to select polymers that delay nucleation and/or API crystallization. Figure S2. The UV-vis quantification method was specific for NTZ at 435 nm. At 2.97 µg/mL of pure NTZ and both cocrystals, there is no interference by coformers in the quantification method used herein. At other concentration levels of NTZ and coformers, the observation was the same, there was no interference in NTZ quantitation by the coformers presence. Figure S3. UV-vis spectra of samples from powders dissolution of NTZ (left) and NTZ-SUC cocrystal (right) formulated at different concentrations of Methocel® 60 HG. NTZ spectrum did not change in presence of polymer. Figure S4. Comparison of PXRD patterns: A. (a) Nitazoxanide, (b) glutaric acid, (c) NTZ-GLU simulated from SCXRD data, (d–m) Batches No. 1–10 of NTZ-GLU cocrystals prepared in l gram scale by SDG with acetone as solvent. B. (a) Nitazoxanide, (b) succinic acid, (c) NTZ-SUC simulated from SCXRD data, (d–m) Batches No. 1–10 of NTZ-SUC cocrystals prepared in l gram scale by SDG with acetone as solvent. Figure S5. SEM images of the solid starting materials used for the powder dissolution tests: (a) NTZ, (b) physical mixture of NTZ and SUC; and solid residues recovered after the dissolution tests of the solids formulated with Methocel® 60 HG: (c) NTZ with 1.0% *w/w*, (d) physical mixture of NTZ and SUC with 1.0% *w/w*, (e) NTZ with 5.0% *w/w*, (f) physical mixture of NTZ and SUC with 5.0% *w/w*. Note: The SEM images of the starting materials are presented at higher resolution than the remaining images (see scale in each image).

Author Contributions: Conceptualization, H.H. (solid-state), H.M.-R. (physical organic) and D.H.-R. (pharmaceutics); methodology, R.S.-Z., C.R.-R., and P.R.-C.; formal analysis and investigation, R.S.-Z.; resources, H.H. and D.H.-R.; writing—original draft preparation, R.S.-Z., H.H., and D.H.-R.; writing—review and editing, H.M.-R.; supervision, O.S.-G.; project administration, D.H.-R., and funding acquisition, H.M.-R. and D.H.-R. All authors approved the final version of this article. All authors have read and agreed to the published version of the manuscript.

Funding: This work received support from Consejo Nacional de Ciencia y Tecnología (CONACyT) and Secretaría de Educación Pública (SEP-PRODEP) in form of postgraduate fellowships for R.S.-Z and C.R.-R. and a postdoctoral fellowship for O.S.-G. Financial support through CONACYT grants No. 404178, INF-2015-251898 and CB-2013-221451 is gratefully acknowledged.

Conflicts of Interest: The authors declare no conflict of interest.

References

1. Rossignol, J.F. Nitazoxanide: A first-in-class broad-spectrum antiviral agent. *Antivir. Res.* **2014**, *110*, 94–103. [CrossRef] [PubMed]

2. Rossignol, J.F. Nitazoxanide, a new drug candidate for the treatment of Middle East respiratory syndrome coronavirus. *J. Infect. Public Health* **2016**, *9*, 227–230. [CrossRef]

3. Shalan, S.; Nasr, J.J.; Belal, F. Determination of tizoxanide, the active metabolite of nitazoxanide, by micellar liquid chromatography using a monolithic column. Application to pharmacokinetic studies. *Anal. Methods* **2014**, *6*, 8682–8689. [CrossRef]

4. Di Santo, N.; Ehrisman, J. A functional perspective of nitazoxanide as a potential anticancer drug. *Mutat. Res. Fundam. Mol. Mech. Mutagen.* **2014**, *768*, 16–21. [CrossRef]

5. Hong, S.K.; Kim, H.J.; Song, C.S.; Choi, I.S.; Lee, J.B.; Park, S.Y. Nitazoxanide suppresses IL-6 production in LPS-stimulated mouse macrophages and TG-injected mice. *Int. Immunopharmacol.* **2012**, *13*, 23–27. [CrossRef] [PubMed]

6. Jasenosky, L.D.; Cadena, C.; Mire, C.E.; Borisevich, V.; Haridas, V.; Ranjbar, S.; Nambu, A.; Bavari, S.; Soloveva, V.; Sadukhan, S. The FDA-Approved Oral Drug Nitazoxanide Amplifies Host Antiviral Responses and Inhibits Ebola Virus. *iScience* **2019**, *19*, 1279–1290. [CrossRef] [PubMed]

7. Ai, N.; Wood, R.D.; Welsh, W.J. Identification of Nitazoxanide as a Group I Metabotropic Glutamate Receptor Negative Modulator for the Treatment of Neuropathic Pain: An in Silico Drug Repositioning Study. *Pharm. Res.* **2015**, *32*, 2798–2807. [CrossRef] [PubMed]

8. Félix-Sonda, B.C.; Rivera-Islas, J.; Herrera-Ruiz, D.; Morales-Rojas, H.; Höpfl, H. Nitazoxanide Cocrystals in Combination with Succinic, Glutaric, and 2,5-Dihydroxybenzoic Acid. *Cryst. Growth Des.* **2014**, *14*, 1086–1102. [CrossRef]

9.   Matysiak-Budnik, T.; Mégraud, F.; Heyman, M. In-vitro transfer of nitazoxanide across the intestinal epithelial barrier. *J. Pharm. Pharmacol.* **2002**, *54*, 1413–1417. [CrossRef]

10.   Duggirala, N.K.; Perry, M.L.; Almarsson, Ö.; Zaworotko, M.J. Pharmaceutical cocrystals: Along the path to improved medicines. *Chem. Commun.* **2016**, *52*, 640–655. [CrossRef]

11.   Bolla, G.; Nangia, A. Pharmaceutical cocrystals: Walking the talk. *Chem. Commun.* **2016**, *52*, 8342–8360. [CrossRef]

12.   Wouters, J.; Quéré, L. *Pharmaceutical Salts and Co-Crystals*; The Royal Society of Chemistry: Cambridge, UK, 2012; pp. 351–371.

13.   Kuminek, G.; Cao, F.; Rocha, A.B.O.; Cardoso, S.G.; Rodriguez-Hornedo, N. Cocrystals to facilitate delivery of poorly soluble compounds beyond-rule-of-5. *Adv. Drug Deliv. Rev.* **2016**, *101*, 143–166. [CrossRef] [PubMed]

14.   Berry, D.J.; Steed, J.W. Pharmaceutical cocrystals, salts and multicomponent systems; intermolecular interactions and property based design. *Adv. Drug Deliv. Rev.* **2017**, *117*, 3–24. [CrossRef]

15.   Sathisaran, I.; Dalvi, S. Engineering Cocrystals of Poorly Water-Soluble Drugs to Enhance Dissolution in Aqueous Medium. *Pharmaceutics* **2018**, *10*, 108. [CrossRef]

16.   Good, D.J.; Rodríguez-Hornedo, N. Solubility Advantage of Pharmaceutical Cocrystals. *Cryst. Growth Des.* **2009**, *9*, 2252–2264. [CrossRef]

17.   Cavanagh, K.L.; Maheshwari, C.; Rodríguez-Hornedo, N. Understanding the Differences Between Cocrystal and Salt Aqueous Solubilities. *J. Pharm. Sci.* **2018**, *107*, 113–120. [CrossRef]

18.   Cao, F.; Rodriguez-Hornedo, N.; Amidon, G.E. Mechanistic Analysis of Cocrystal Dissolution, Surface pH, and Dissolution Advantage as a Guide for Rational Selection. *J. Pharm. Sci.* **2019**, *108*, 243–251. [CrossRef] [PubMed]

19.   Huang, Y.; Kuminek, G.; Roy, L.; Cavanagh, K.L.; Yin, Q.; Rodríguez-Hornedo, N. Cocrystal Solubility Advantage Diagrams as a Means to Control Dissolution, Supersaturation, and Precipitation. *Mol. Pharm.* **2019**, *16*, 3887–3895. [CrossRef] [PubMed]

20.   Brouwers, J.; Brewster, M.E.; Augustijns, P. Supersaturating Drug Delivery Systems: The Answer to Solubility-Limited Oral Bioavailability? *J. Pharm. Sci.* **2009**, *98*, 2549–2572. [CrossRef] [PubMed]

21.   Almeida e Sousa, L.; Reutzel-Edens, S.M.; Stephenson, G.A.; Taylor, L.S. Supersaturation Potential of Salt, Co-Crystal, and Amorphous Forms of a Model Weak Base. *Cryst. Growth Des.* **2016**, *16*, 737–748. [CrossRef]

22.   Xu, S.; Dai, W.G. Drug precipitation inhibitors in supersaturable formulations. *Int. J. Pharm.* **2013**, *453*, 36–43. [CrossRef] [PubMed]

23.   Remenar, J.F.; Peterson, M.L.; Stephens, P.W.; Zhang, Z.; Zimenkov, Y.; Hickey, M.B. Celecoxib: Nicotinamide Dissociation: Using Excipients To Capture the Cocrystal's Potential. *Mol. Pharm.* **2007**, *4*, 386–400. [CrossRef] [PubMed]

24.   Childs, S.L.; Kandi, P.; Lingireddy, S.R. Formulation of a Danazol Cocrystal with Controlled Supersaturation Plays an Essential Role in Improving Bioavailability. *Mol. Pharm.* **2013**, *10*, 3112–3127. [CrossRef] [PubMed]

25.   Alhalaweh, A.; Ali, H.R.H.; Velaga, S.P. Effects of Polymer and Surfactant on the Dissolution and Transformation Profiles of Cocrystals in Aqueous Media. *Cryst. Growth Des.* **2014**, *14*, 643–648. [CrossRef]

26.   Li, M.; Qiu, S.; Lu, Y.; Wang, K.; Lai, X.; Rehan, M. Investigation of the Effect of Hydroxypropyl Methylcellulose on the Phase Transformation and Release Profiles of Carbamazepine-Nicotinamide Cocrystal. *Pharm. Res.* **2014**, *31*, 2312–2325. [CrossRef] [PubMed]

27.   Ullah, M.; Ullah, H.; Murtaza, G.; Mahmood, Q.; Hussain, I. Evaluation of Influence of Various Polymers on Dissolution and Phase Behavior of Carbamazepine-Succinic Acid Cocrystal in Matrix Tablets. *BioMed Res. Int.* **2015**, *2015*, 10. [CrossRef] [PubMed]

28.   Ullah, M.; Hussain, I.; Sun, C.C. The development of carbamazepine-succinic acid cocrystal tablet formulations with improved in vitro and in vivo performance. *Drug Dev. Ind. Pharm.* **2016**, *42*, 969–976. [CrossRef]

29.   Qiu, S.; Li, M. Effects of coformers on phase transformation and release profiles of carbamazepine cocrystals in hydroxypropyl methylcellulose based matrix tablets. *Int. J. Pharm.* **2015**, *479*, 118–128. [CrossRef]

30.   Qiu, S.; Lai, J.; Guo, M.; Wang, K.; Lai, X.; Desai, U.; Juma, N.; Li, M. Role of polymers in solution and tablet-based carbamazepine cocrystal formulations. *CrystEngComm* **2016**, *18*, 2664–2678. [CrossRef]

31.   Li, M.; Qiao, N.; Wang, K. Influence of Sodium Lauryl Sulfate and Tween 80 on Carbamazepine–Nicotinamide Cocrystal Solubility and Dissolution Behaviour. *Pharmaceutics* **2013**, *5*, 508–524. [CrossRef]

32.   Guo, M.; Wang, K.; Hamill, N.; Lorimer, K.; Li, M. Investigating the Influence of Polymers on Supersaturated Flufenamic Acid Cocrystal Solutions. *Mol. Pharm.* **2016**, *13*, 3292–3307. [CrossRef] [PubMed]

33. Yoshimura, M.; Miyake, M.; Kawato, T.; Bando, M.; Toda, M.; Kato, Y.; Fukami, T.; Ozeki, T. Impact of the Dissolution Profile of the Cilostazol Cocrystal with Supersaturation on the Oral Bioavailability. *Cryst. Growth Des.* **2017**, *17*, 550–557. [CrossRef]

34. He, H.; Zhang, Q.; Li, M.; Wang, J.R.; Mei, X. Modulating the Dissolution and Mechanical Properties of Resveratrol by Cocrystallization. *Cryst. Growth Des.* **2017**, *17*, 3989–3996. [CrossRef]

35. Wang, C.; Tong, Q.; Hou, X.; Hu, S.; Fang, J.; Sun, C.C. Enhancing Bioavailability of Dihydromyricetin through Inhibiting Precipitation of Soluble Cocrystals by a Crystallization Inhibitor. *Cryst. Growth Des.* **2016**, *16*, 5030–5039. [CrossRef]

36. Shimpi, M.R.; Alhayali, A.; Cavanagh, K.L.; Rodríguez-Hornedo, N.; Velaga, S.P. Tadalafil–Malonic Acid Cocrystal: Physicochemical Characterization, pH-Solubility, and Supersaturation Studies. *Cryst. Growth Des.* **2018**, *18*, 4378–4387. [CrossRef]

37. Jasani, M.S.; Kale, D.P.; Singh, I.P.; Bansal, A.K. Influence of Drug–Polymer Interactions on Dissolution of Thermodynamically Highly Unstable Cocrystal. *Mol. Pharm.* **2019**, *16*, 151–164. [CrossRef]

38. Kuminek, G.; Cavanagh, K.L.; da Piedade, M.F.M.; Rodríguez-Hornedo, N. Posaconazole Cocrystal with Superior Solubility and Dissolution Behavior. *Cryst. Growth Des.* **2019**, *19*, 6592–6602. [CrossRef]

39. Sun, D.D.; Wen, H.; Taylor, L.S. Non-Sink Dissolution Conditions for Predicting Product Quality and in Vivo Performance of Supersaturating Drug Delivery Systems. *J. Pharm. Sci.* **2016**, *105*, 2477–2488. [CrossRef]

40. Chen, Y.; Li, L.; Yao, J.; Ma, Y.Y.; Chen, J.M.; Lu, T.B. Improving the Solubility and Bioavailability of Apixaban via Apixaban–Oxalic Acid Cocrystal. *Cryst. Growth Des.* **2016**, *16*, 2923–2930. [CrossRef]

41. Ferretti, V.; Dalpiaz, A.; Bertolasi, V.; Ferraro, L.; Beggiato, S.; Spizzo, F.; Spisni, E.; Pavan, B. Indomethacin Co-Crystals and Their Parent Mixtures: Does the Intestinal Barrier Recognize Them Differently? *Mol. Pharm.* **2015**, *12*, 1501–1511. [CrossRef]

42. Liu, M.; Hong, C.; Yao, Y.; Shen, H.; Ji, G.; Li, G.; Xie, Y. Development of a pharmaceutical cocrystal with solution crystallization technology: Preparation, characterization, and evaluation of myricetin-proline cocrystals. *Eur. J. Pharm. Biopharm.* **2016**, *107*, 151–159. [CrossRef] [PubMed]

43. Weyna, D.R.; Cheney, M.L.; Shan, N.; Hanna, M.; Zaworotko, M.J.; Sava, V.; Song, S.; Sanchez-Ramos, J.R. Improving Solubility and Pharmacokinetics of Meloxicam via Multiple-Component Crystal Formation. *Mol. Pharm.* **2012**, *9*, 2094–2102. [CrossRef] [PubMed]

44. Suresh, K.; Mannava, M.C.; Nangia, A. Cocrystals and alloys of nitazoxanide: Enhanced pharmacokinetics. *Chem. Commun.* **2016**, *52*, 4223–4226. [CrossRef] [PubMed]

45. Rodriguez-Morales, A.J.; Martinez-Pulgarin, D.F.; Muñoz-Urbano, M.; Gómez-Suta, D.; Sánchez-Duque, J.A.; Machado-Alba, J.E. Bibliometric Assessment of the Global Scientific Production of Nitazoxanide. *Cureus* **2017**, *9*, 1204. [CrossRef]

46. Hotez, P.J. Could Nitazoxanide Be Added to Other Essential Medicines for Integrated Neglected Tropical Disease Control and Elimination? *PLOS Negl. Trop. Dis.* **2014**, *8*, 2758. [CrossRef]

47. Aronson, H. Correction Factor for Dissolution Profile Calculations. *J. Pharm. Sci.* **1993**, *82*, 1190. [CrossRef]

48. Guzmán, H.R.; Tawa, M.; Zhang, Z.; Ratanabanangkoon, P.; Shaw, P.; Gardner, C.R.; Chen, H.; Moreau, J.P.; Almarsson, Ö.; Remenar, J.F. Combined Use of Crystalline Salt Forms and Precipitation Inhibitors to Improve Oral Absorption of Celecoxib from Solid Oral Formulations. *J. Pharm. Sci.* **2007**, *96*, 2686–2702. [CrossRef]

49. Chen, Y.M.; Rodríguez-Hornedo, N. Cocrystals Mitigate Negative Effects of High pH on Solubility and Dissolution of a Basic Drug. *Cryst. Growth Des.* **2018**, *18*, 1358–1366. [CrossRef]

50. Kavanagh, O.N.; Croker, D.M.; Walker, G.M.; Zaworotko, M.J. Pharmaceutical cocrystals: From serendipity to design to application. *Drug Discov. Today* **2019**, *24*, 796–804. [CrossRef]

51. Kale, D.P.; Zode, S.S.; Bansal, A.K. Challenges in Translational Development of Pharmaceutical Cocrystals. *J. Pharm. Sci.* **2017**, *106*, 457–470. [CrossRef]

52. Warren, D.B.; Benameur, H.; Porter, C.J.; Pouton, C.W. Using polymeric precipitation inhibitors to improve the absorption of poorly water-soluble drugs: A mechanistic basis for utility. *J. Drug Target.* **2010**, *18*, 704–731. [CrossRef] [PubMed]

53. Alonzo, D.E.; Raina, S.; Zhou, D.; Gao, Y.; Zhang, G.G.; Taylor, L.S. Characterizing the Impact of Hydroxypropylmethyl Cellulose on the Growth and Nucleation Kinetics of Felodipine from Supersaturated Solutions. *Cryst. Growth Des.* **2012**, *12*, 1538–1547. [CrossRef]

54. Ilevbare, G.A.; Liu, H.; Edgar, K.J.; Taylor, L.S. Impact of Polymers on Crystal Growth Rate of Structurally Diverse Compounds from Aqueous Solution. *Mol. Pharm.* **2013**, *10*, 2381–2393. [CrossRef] [PubMed]

55. Shiraki, K.; Takata, N.; Takano, R.; Hayashi, Y.; Terada, K. Dissolution Improvement and the Mechanism of the Improvement from Cocrystallization of Poorly Water-soluble Compounds. *Pharm. Res.* **2008**, *25*, 2581–2592. [CrossRef] [PubMed]
56. Jayasankar, A.; Reddy, L.S.; Bethune, S.J.; Rodríguez-Hornedo, N. Role of Cocrystal and Solution Chemistry on the Formation and Stability of Cocrystals with Different Stoichiometry. *Cryst. Growth Des.* **2009**, *9*, 889–897. [CrossRef]
57. Babu, N.J.; Nangia, A. Solubility Advantage of Amorphous Drugs and Pharmaceutical Cocrystals. *Cryst. Growth Des.* **2011**, *11*, 2662–2679. [CrossRef]
58. Yalkowsky, S.H.; He, Y.; Jain, P. *HANDBOOK OF Aqueous Solubility Data*; CRC Press: Boca Raton, FL, USA, 2010; pp. 96–154.

Article

The Role of Cocrystallization-Mediated Altered Crystallographic Properties on the Tabletability of Rivaroxaban and Malonic Acid

Dnyaneshwar P. Kale [1], Vibha Puri [2], Amit Kumar [3], Navin Kumar [3] and Arvind K. Bansal [1,*]

[1] Department of Pharmaceutics, National Institute of Pharmaceutical Education and Research (NIPER), S.A.S. Nagar 160062, India; kale_pep15@niper.ac.in
[2] Bristol Myers Squibb, 556 Morris Avenue, Summit, NJ 07901, USA; puri.vibhaa@gmail.com
[3] Department of Mechanical Engineering, Indian Institute of Technology (IIT) Ropar, Rupnagar 140001, India; amitkamboj310@gmail.com (A.K.); nkumar@iitrpr.ac.in (N.K.)
* Correspondence: akbansal@niper.ac.in; Tel.: +91-172-2214682-2126; Fax: +91-172-221-4692

Received: 16 May 2020; Accepted: 8 June 2020; Published: 12 June 2020

Abstract: The present work aims to understand the crystallographic basis of the mechanical behavior of rivaroxaban-malonic acid cocrystal (RIV-MAL Co) in comparison to its parent constituents, i.e., rivaroxaban (RIV) and malonic acid (MAL). The mechanical behavior was evaluated at the bulk level by performing "out of die" bulk compaction and at the particle level by nanoindentation. The tabletability order for the three solids was MAL < RIV < RIV-MAL Co. MAL demonstrated "lower" tabletability because of its lower plasticity, despite it having reasonably good bonding strength (BS). The absence of a slip plane and "intermediate" BS contributed to this behavior. The "intermediate" tabletability of RIV was primarily attributed to the differential surface topologies of the slip planes. The presence of a primary slip plane (0 1 1) with flat-layered topology can favor the plastic deformation of RIV, whereas the corrugated topology of secondary slip planes (1 0 $\bar{2}$) could adversely affect the plasticity. In addition, the higher elastic recovery of RIV crystal also contributed to its tabletability. The significantly "higher" tabletability of RIV-MAL Co among the three molecular solids was the result of its higher plasticity and BS. Flat-layered topology slip across the (0 0 1) plane, the higher degree of intermolecular interactions, and the larger separation between adjacent crystallographic layers contributed to improved mechanical behavior of RIV-MAL Co. Interestingly, a particle level deformation parameter H/E (i.e., ratio of mechanical hardness H to elastic modulus E) was found to inversely correlate with a bulk level deformation parameter σ_0 (i.e., tensile strength at zero porosity). The present study highlighted the role of cocrystal crystallographic properties in improving the tabletability of materials.

Keywords: cocrystal; compaction; nanoindentation; slip plane; tabletability; surface topology; interparticulate bonding area; interparticulate bonding strength

---

1. Introduction

Oral solid dosage forms (such as tablets and capsules) represent the most popular delivery system amongst various pharmaceutical dosage forms, as they offer many benefits in terms of cost, stability, ease of handling, and patient compliance [1]. The mechanical properties of active pharmaceutical ingredients (APIs) play a critical role in the manufacturing of solid dosage form, as poor mechanical properties can cause difficulties during processing. With the emergence of crystal engineering, cocrystallization has become a useful strategy in the design of pharmaceutical materials with desired properties [2–5]. Cocrystallization has received significant attention because it can be employed for altering physicochemical properties of the APIs such as solubility [6], dissolution [7,8], stability [9], hygroscopicity [10,11], and mechanical properties [12,13]. Many reports suggest that cocrystallization can improve [14,15], deteriorate [16,17] or have no impact on modulation of mechanical properties of organic molecular solids including APIs. Understanding the role of crystallographic/supramolecular features, in modulating mechanical properties, remains an area of significant interest.

Tabletability is represented by tensile strength as a function of applied compaction pressure. It has been commonly used to compare the compaction performance of organic molecular solids and is particularly useful in studying bulk deformation behavior [18,19]. The tensile strength of the compacts is governed by interparticulate bonding area (BA) as well as interparticulate bonding strength (BS) [20]. Therefore, a thorough understanding of tabletability can be developed by a quantitative model, based on the concept of interparticulate bonding area and bonding strength (BABS model) [21,22]. Increased BA is achieved on irreversible (plastic) deformation of organic molecular solid and this deformation behavior is known as plasticity [21]. Crystallographic features such as slip systems are responsible for plasticity (increase in BA) in single component polymorphs [19,23] and multicomponent cocrystals [14,24]. The degree of intermolecular interactions and true density contribute to the BS of the crystal form [19,20,25,26]. It is worthwhile mentioning that "plasticity" only contributes to BA, and hence the role of BS also needs to be considered while evaluating bulk deformation (compaction) behavior.

Nano-mechanical parameters such as elastic modulus (E), indentation hardness (H) and elastic recovery (determined from load-displacement (P–h) curves) provide useful insights on the mechanical behavior of solids at a particle (crystal) level [27–29]. It is noteworthy that nanoindentation focuses on particle mechanical properties, but other factors such as particle size, particle shape and distribution, inter-particle interaction can have a significant effect when translating these properties to the bulk deformation (compaction). Interestingly, nanoindentation is useful for interpolation of particle-level behavior to crystallographic features [30], thereby furthering our understanding of crystal structure-mechanical property relationships. As demonstrated by Wendy and Duncan-Hewitt [31], and recently by our research group [32], the indentation hardness to elastic modulus (H/E) ratio is a predictor of compaction behaviour of materials undergoing deformation. The higher the value of H/E, the higher are the residual stresses after compaction (higher elastic recovery) and vice-versa. Another predictor of elastic recovery is (h_{max}-h_p), wherein h_p is the permanent indentation depth after removal of the test load and h_{max} is indentation depth at a maximum load; both are computed from the (P–h) curve.

Crystal anisotropy is responsible for distinct variations of the plastic deformation within the same structural class. Consequently, deformation processes in molecular solids cannot be labelled in simple terms and each molecular solid has to be evaluated individually [29,33]. It is well accepted that the presence of active slip planes in molecular solids leads to "plastic" deformation. A higher plasticity confers improved tabletability, because of increased BA. Therefore, slip planes have been increasingly used to describe "plasticity" and have also been used to correlate to the deformation behavior of molecular solids. However, the existing literature lacks descriptions of how a molecular solid undergoes deformation when its crystal structure contains more than one active slip system with two different surface topologies (flat and zigzag topologies).

In the present work, we reveal the role of crystallographic features such as slip planes, slip plane topology, intermolecular interactions (nature of hydrogen bonding: 1D, 2D, or 3D) and largest d-spacing on "plasticity" and interparticulate bonding strength of the model systems—rivaroxaban (RIV), malonic acid (MAL) and 2:1 rivaroxaban-malonic acid cocrystal (RIV-MAL Co). In particular, we focus on how the materials deform when their crystal structures contain a single active slip plane system (RIV-MAL Co), more than one slip plane system (RIV) and devoid of active slip planes (MAL). In this study, the nanoindentation technique was employed to uncover "crystal level" deformation behavior, while "bulk level" deformation was studied using Compressibility-Tabletability-Compactibility (CTC) profiling. Slip plane systems were predicted by visualization as well as attachment energy calculations, and further experimentally confirmed by nanoindentation studies.

2. Materials

Molecular structures of RIV and MAL are presented in Figure 1. RIV (Purity > 99.5%) was generously provided as a gift sample by MSN Laboratories, India. Malonic acid (MAL) and 2,2,2-trifluoroethanol were purchased from Sigma-Aldrich, St. Louis, MO, USA and Spectrochem Pvt. Ltd., Mumbai, India, respectively, and used as received. Crystallization experiments were carried out to generate samples for compaction. For nanoindentation experiments, relatively larger-sized crystals were generated, and the crystallization methodology for same has been thoroughly discussed in our recent publication [10].

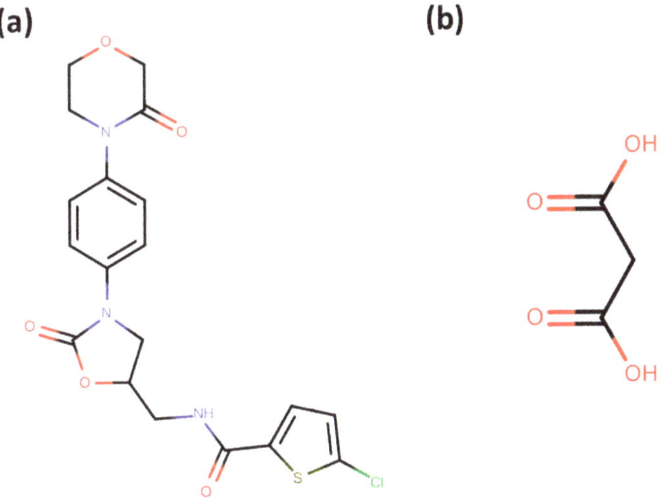

Figure 1. Molecular Structure of (**a**) RIV and (**b**) MAL.

3. Methods

3.1. Crystallization Experiments

3.1.1. RIV for Compaction Study

A saturated solution of RIV was prepared by dissolving the drug (~6.0 g) in acetonitrile (600 mL) with magnetic stirring the solution on a heated water bath at 60-65 °C for 30 min followed by cooling at room temperature (RT). The resultant solution was kept for slow solvent evaporation at room temperature. Columnar-shaped, small-sized crystals, generated within 1-2 h, were collected by vacuum filtration and subjected to air drying.

3.1.2. MAL for Compaction Study

Crystalline MAL from the supplier was ground using a mortar and pestle to obtain the desired particle size range of 5-40 μm. Prior to the compaction study, the powder was dried in a hot air oven at 50 ± 2 °C for 2 h to remove any adsorbed moisture from the surface.

3.1.3. RIV-MAL Co for Compaction Study

The compaction study requires a large quantity of material; hence the previously reported process was scaled up. Initially, a small batch (~150 mg) of RIV-MAL Co was prepared as per the reported method [10]. These crystals were used as "seeds" for preparing a batch of ~8.0 g.

For preparing 8.0 g batch of RIV-MAL Co, 6.0 g of RIV was dissolved in 30.1 mL of 2,2,2-trifluoroethanol (TFE) in a glass beaker (50 mL capacity) with stirring and heated using a water bath between 70 and 80 °C. MAL (2.07 g) was added to the hot solution of RIV and volume of the solution was reduced to approximately 55% by heating (70-75 °C) it in the open vessel for 1-2 h. Thereafter, the hot solution was transferred to another glass beaker. The resultant solution was seeded with the crystals of RIV-MAL Co and cooled to room temperature. After 10-20 min, the cocrystals crystallized from the solution were collected using vacuum filtration. The crystals were dried at 25 °C in a vacuum drier.

All three solids were stored in a desiccator at 43% RH (maintained using a saturated solution of potassium carbonate, K_2CO_3) until further use.

3.2. Differential Scanning Calorimetry (DSC)

The melting point and heat of fusion of samples were determined using DSC instrument (model Q2000, TA Instruments, New Castle, DE, USA) equipped with a refrigerated cooling system. The instrument was operating with TA Universal Analysis® software (version 4.5 A) and calibrated initially for temperature and heat flow using high purity standard of indium. The sample (2–4 mg) was analyzed in a T_{zero} aluminium pan crimped with a lid. The weight of sample was accurately recorded on a microbalance (Sartorius GM 1202, Goettingen, Germany). Each sample was subjected to thermal scanning at a heating rate of 10.0 °C/min. The purge of dry nitrogen at flow rate of 50 mL/min was maintained during analysis.

3.3. X-Ray Powder Diffractometry (XRPD)

XRPD of the samples was collected at room temperature (25.0 ± 2.0 °C) on a Bruker D8 Advance Diffractometer (Bruker, AXS, Karlsruhe, Germany). The instrument was operating with Cu Kα radiation (1.54 Å) passing through a nickel filter, and the tube voltage was set at 40.0 kV and current of the generator (amperage) at 40.0 mA. The sample (250-300 mg) was mounted on a poly(methyl methacrylate) (PMMA) sample holder and pressed with a glass slide to achieve coplanarity with the surface of sample holder. Each sample was then scanned in 2θ range of 5.0°-40.0° with a step size of 0.05° and a step time of 1.0 s. The obtained data were plotted using OriginPro software (OriginLab Corporation, Northampton, MA, USA).

3.4. Particle Size Distribution (PSD)

Comparable particle sizes were obtained by passing the samples through a British sieve size (BSS) No. 100# and retaining on a 120# sieve. Each sample was mounted on a glass slide with silicone oil and the observed under optical microscope (Leica DMLP, Leica Microsystems, Wetzlar, Germany) for measuring particle diameter. The diameter (i.e., length along the longest axis of individual particles) of 300 particles were measured. Particle size distribution curve was plotted to determine the diameters corresponding to 10%, 50%, and 90% of cumulative undersize particles, i.e., D_{10}, D_{50}, and D_{90}.

3.5. Specific Surface Area (SSA) Measurement

The specific surface area of RIV, MAL and RIV-MAL Co was determined by nitrogen gas adsorption using "Smart Sorb 92" surface area analyzer (Smart Instruments, Mumbai, India). The gas loop of the instrument was filled with known amount of sample (500–800 mg) and then submerged into liquid nitrogen for adsorption. After completion of the adsorption cycle, a desorption was triggered by submerging the glass loop into water at ambient condition. The quantity of the adsorbed gas was measured using a thermal conductivity detector and the data were integrated using an electronic circuit in terms of counts. The measured parameters were then used to calculate the surface area of the sample by employing the adsorption theories of Brunauer, Emmett, and Teller. SSA was reported as average of three individual measurements (n = 3).

3.6. Moisture Content (MC)

Karl Fischer titration using "Titrino 794 Basic titrator" (Metrohm AG, Herisau, Switzerland) was used to measure moisture content (water content) of samples. Prior to analysis, calibration of the instrument was performed using disodium tartrate dehydrate. MC was reported as the average of three measurement for each sample.

3.7. True Density Determination

The true density of samples was determined using a helium gas pycnometer (Pycno 30, Smart Instruments, Mumbai, India) as per the previously reported protocol [34]. Before analysis, the samples were pre-dried at 40 °C for 24 h in a vacuum oven to avoid the effect of residual moisture on true density measurement. The pre-dried sample, sufficient to fill 3/4 of the volume of the sample cell, was weighed (1.5–2.5 g) and transferred into the sample cell. The first pressure reading (P_1) was recorded after passing a pressurized pure helium gas in a known reference volume into the reference cell. Then, the pressurized helium gas was allowed to flow from the reference cell into the sample cell. This led to drop in the initial pressure that recorded as second pressure reading (P_2). These values of P1 and P2 put into Equation (1) to calculate the true volume V_P.

$$V_P = \left(\frac{P_1}{P_2} - 1 \right)(V_c - V_r) \tag{1}$$

where V_c and V_r are the cell volume and reference volume having values of 18.9522 and 11.9587 cm^3/g, respectively. True density was calculated by dividing the sample mass by true volume (V_P) value.

3.8. Preparation of Compacts for Studying Bulk Deformation Behavior

Compacts were prepared by compressing 400 mg of crystalline powder using different compaction pressure in a hydraulic press (Type KP, Sr. No. 1125, Kimaya Engineers, Maharastra, India). The applied dwell time for compaction preparation was 1.0 min using a 13.0 mm punch die set (round, flat-faced punch). Different compression forces were applied manually to achieve a range of compaction pressures from 37-222 MPa. The actual compaction pressure was determined from the know value of the applied hydraulic load (Force) and the surface area of the flat punch-die set used for compression. Equation (2) was used for converting the applied load into compaction pressures.

$$P = \frac{F}{A} \tag{2}$$

where, F is the applied hydraulic load (Newton, N), and A is the surface area of the flat punch-die set (in mm^2). Prior to analysis, the prepared compacts were stored for 48 h under ambient conditions to allow for relaxation of any residual stress. Subsequently, compacts were analyzed for weight, thickness, and breaking force measurement.

3.9. Calculation of Tensile Strength and Porosity

The assessment of tensile strength (σ) value in bulk deformation profiling helps to eliminate the undesirable effect of variable tablet thickness on a measured breaking force. Therefore, tensile strength (σ) was calculated using Equation (3), based on breaking force (F), table diameter (d) and tablet thickness (t).

$$\sigma = \frac{2F}{\pi dt} \tag{3}$$

where σ is the tensile strength in MPa, F is the observed breaking force in N, d is the diameter in mm, and t is thickness of the compact in mm.

Tablet hardness tester (Erweka, TBH 20, USA) was used to measure the breaking force (F) of all of the compacts. Digital caliper (CD-6 CS, Digimatic Mitutoyo Corporation, Kanagawa, Japan) was used to measure tablet diameter and thickness. The porosity (ε) of the compacts was calculated using Equation (4), from tablet density (ρ_c) and true density of the powder (ρ_t). ρ_c is calculated from the weight and volume of the resulting tablet, while ρ_t is measured by helium pycnometer as described above.

$$\varepsilon = 1 - \frac{\rho_c}{\rho_t} \tag{4}$$

3.10. Nanoindentation Experimentation

Nanoindentation was performed on oriented single crystals of RIV, MAL and RIV-MAL Co using a Ti-950 TriboIndenter (Hysitron Inc., Minneapolis, MN, USA) using a protocol, which is similar to that of previously reported by our research group [32]. Briefly, the indenter employed was a tripyramidal (Berkovich) tip having an inclined angle of 142.3° and a tip radius of ~150 nm. The fused silica and polycarbonate standards were used to calibrate the tip area function. The tip area function calibration was carried out by performing a series of indents with different contact depths on a standard sample of known elastic modulus (E). A plot of the calculated area against contact depth (h) was created and fitted by the TriboScan software. An optical microscope integrated into the nanoindentation system was used to identify the regions on crystal surface for testing. The "tip to optics calibration" was undertaken by performing 10 indents in an "H-pattern". The testing was carried out at 28 ± 0.5 °C temperature and 45 ± 5% relative humidity. For quasi-static analysis of all samples, 10–12 subsequent indents were performed along the length, midline parallel to the longest axis of the crystal on a dominant face with user-specified parameters. The sufficient contact depths, large enough to local surface roughness were estimated to avoid strong effect of roughness on the measured mechanical properties. The peak load (P) for these indentations was 1000 μN, and the indent spacing was 55.0 μm. A load function consisting of a 5 s loading to peak force (F) segment, followed by a 2-s hold segment and a 5-s unloading segment was used (the loading and unloading rates were 0.2 mN/s). The Oliver and Pharr method [35] was employed to compute the nanomechanical hardness (H) and the elastic modulus (E_r). The E_r value is related to the Young modulus of elasticity of the tested sample (E_s) and the indenter (E_i) through the following relationship in Equation (5):

$$\frac{1}{E_r} = \frac{\left(1 - v_i^2\right)}{E_i} + \frac{\left(1 - v_s^2\right)}{E_s} \tag{5}$$

where v_i is Poisson ratio for the indenter material, while v_s is Poisson ratio of the substrate material. The values of the elastic modulus and Poisson ratio for the diamond indenter tip are 1140 GPa and 0.07, respectively.

3.11. Molecular Modeling

The crystal structures of MAL, RIV and RIV-MAL Co were examined for identification of slip planes, d-spacing, intermolecular interactions and H-bonding dimensionalities using Mercury software (Version 3.7, Cambridge Crystallographic Data Centre, CCDC, Cambridge, UK). Previously, our lab has successfully solved and deposited crystal structures of RIV and RIV-MAL Co with the CCDC numbers 1854617 and 1854618, respectively [10]. The CIF file of MAL (CSD Reference code MALNAC02, deposition number 1209218) was downloaded from the CCDC website, https://www.ccdc.cam.ac.uk/structures/.

3.12. Attachment Energy Calculations

The attachment energy, E_{att}, is defined as the energy released on the attachment of a growth slice with the thickness d_{hkl} to a growing crystal face [36,37]. E_{att} is calculated as $E_{att} = E_{lattice} - E_{slice}$, where $E_{lattice}$ is the lattice energy of the crystal, and E_{slice} is the energy released on the formation of a growth slice of a thickness equal to the interplanar d-spacing, d_{hkl}.

Crystal morphology predictions were performed using Material Studio 2018 (Biovia, San Diego, CA, USA). The growth morphology method (with COMPASS II force field) was used to generate E_{att} of different crystal faces. All calculations were performed at fine quality with "Ewald" electrostatic summation method and "atom-based" van der Waals summation method. The crystallographic planes with least total E_{att} were identified as slip planes. Additionally, E_{att} calculations were also performed using Dreiding force field (with "charge Qeq" and "current charge").

3.13. Statistical Analysis

The statistical significance of various bulk deformation parameters was compared using the paired t-test, assuming equal variances using GraphPad Prism 5 software, version 5.04 (GraphPad Software, Inc., San Diego, CA, USA). The test was considered to be statistically significant if $p < 0.05$.

4. Results and Discussion

4.1. Solid-State Characterization

The overlay of DSC heating scans and XRPD patterns for RIV, MAL and RIV-MAL Co are presented in Figures 2 and 3, respectively. The DSC heating curve of the RIV crystals displayed a melting endotherm at 230.5 °C (T_m, onset) with a heat of fusion (ΔH_f) of 118.6 J/g, which matches to the melting temperature of rivaroxaban form I (Figure 2). The XRPD pattern of the generated RIV sample exhibited sharp diffraction peaks at 2θ values of 9.1°, 12.2°, 14.4°, 16.6°, 17.6°, 18.2°, 19.6°, 20.0°, 21.8°, 22.6°, 23.5°, 24.1°, 24.8°, 25.8°, 26.8°, 29.6°, 30.3°, 31.9° ± 0.2° (Figure 3). The obtained diffraction pattern matches to the calculated powder pattern (CCDC identification code—LEMSOO01, deposition no. 1854617) [10] and the peak positions correspond to the values reported for RIV form I. The DSC and XRPD data together confirmed the identity and solid form purity of RIV sample generated for the compaction study.

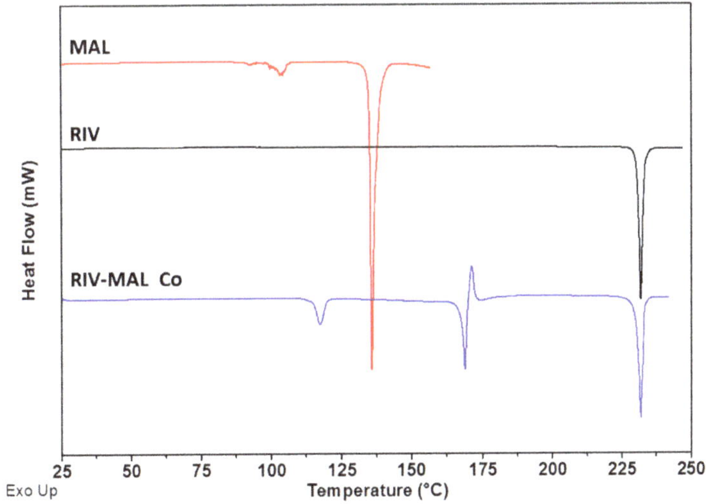

Figure 2. DSC heating curves overlay for RIV, MAL, and RIV-MAL Co Samples.

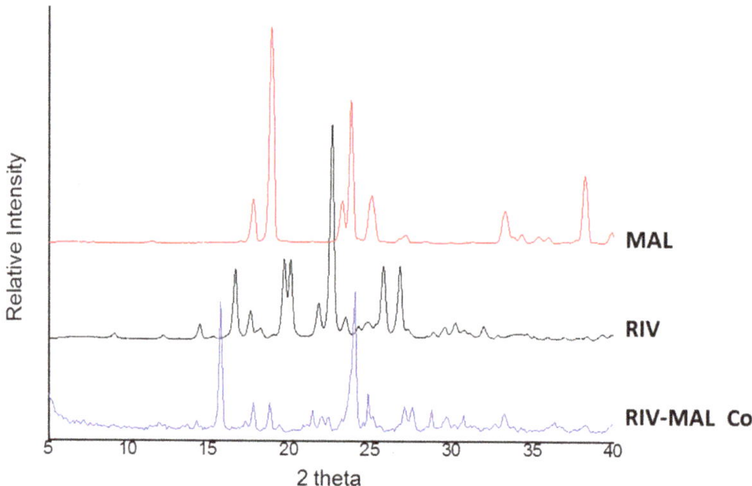

Figure 3. Overlay of XRPD diffractograms for RIV, MAL, and RIV-MAL Co Samples.

The DSC heating curve of MAL sample showed a broad endothermic event between 85.0° to 109.0 °C (related to solid-solid phase transition) followed by a sharp melting endotherm at 135.0 °C (T_m, onset), ($\Delta H_f \approx 235.9$ J/g) [38]. The XRPD pattern of the sample showed sharp diffraction peaks, which correspond to the β form of MAL.

As shown in the DSC analysis of the RIV-MAL Co, the sample showed a transition point between 110 to 122 °C (related to solid to solid phase transition), followed by the endothermic peak corresponding to cocrystal melting at 167.9 °C (ΔH_f is ~36 J/g)[10]. Following the cocrystal melting, the exothermic event (recrystallization of RIV) and the endothermic at ~230 °C (melting of the recrystallized RIV) were observed. The above-thermal events were consistent with the previously reported behaviour of RIV-MAL Co [10]. The experimental diffraction pattern of RIV-MAL Co sample (Figure 3) matched the calculated powder pattern obtained from single-crystal data reported by our laboratory (CCDC identification code- YORVEJ, deposition No. 1854618) [10]. Thus, DSC and XRPD together confirmed a solid form purity of the scale-up batches.

4.2. Particle Level and Bulk Level Attributes

Particle size analysis revealed that all three solids had similar D_{50} and D_{90} values (Table 1). Similar particle size distribution allows better comparison of compaction behavior of these solids at the molecular and supramolecular level. MAL, RIV, and the cocrystal had a moisture content of less than 0.3% *w/w* (Table 1). MAL had a significantly higher true density value (1.628 ± 0.001 g/cm³) as compared to RIV and MAL (1.536 ± 0.005 g/cm³ for RIV and 1.534 ± 0.007 g/cm³).

Table 1. Physical Characterization of MAL, RIV, and RIV-MAL Co.

Material	PSD (µm)			SSA (m²/g)	MC (% *w/w*)	True Density (g/cm³)	
	D_{10}	D_{50}	D_{90}			Experimental	Crystallographic
MAL	8.3	14.2	37.8	0.78 (0.09)	0.264 (0.013)	1.628 (0.001)	1.621
RIV	7.1	10.2	31.5	0.85 (0.04)	0.235 (0.014)	1.536 (0.005)	1.554
RIV-MAL Co	7.8	11.9	35.2	0.83 (0.07)	0.242 (0.012)	1.534 (0.007)	1.548

Values in parentheses indicate standard deviation (SD), n = 3. PSD—Particle size distribution, SSA—Specific surface area, MC—Moisture content.

4.3. Bulk Deformation Behaviour

The deformation behavior of pharmaceutical materials at a bulk level is commonly studied by performing compressibility, tabletability, compactibility (CTC) profiling, as it provides a comprehensive understanding on the role of interparticulate bonding area (BA) and bonding strength (BS) [25,26,39–41]. A measure of the ability of powder material to undergo reduction in volume under the application of compaction pressure is known as "compressibility". Compressibility is represented as a plot of porosity against compaction pressure and the plot signifies the extent of increase in interparticulate bonding area (BA).

The compressibility profiles of the three solids demonstrated greater compressibility of RIV and RIV-MAL Co over MAL at a given compaction pressure (Figure 4a). Slightly higher compressibility of the cocrystal over RIV was observed at higher compaction pressure (>150 MPa) and the differences were statistically significant ($p < 0.05$). The applied compaction pressure may lead to pressure-induced phase transformation in the samples. When examined by using DSC analysis, the compacts of the three molecular solids did not show any evidence of pressure-induced phase transformation.

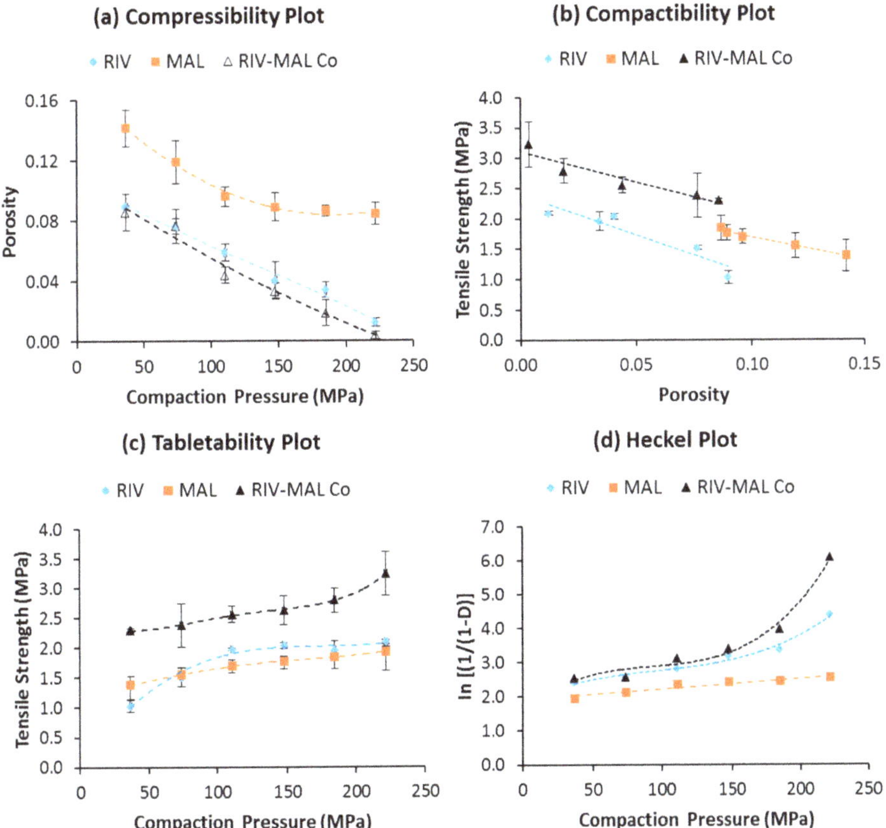

Figure 4. (**a**)–(**d**) Compressibility, tabletability, and compactibility, and Heckel plots for MAL, RIV, and RIV-MAL Co.

Compactibility is defined as "*the ability of the powder material to produce tablets of sufficient tensile strength under the effect of densification*" [21,25]. It is represented by a plot of tensile strength against tablet porosity, and the plot signifies the bonding strength (BS) of a given material. Compactibility plots (Figure 4b) indicate a higher bonding strength of the cocrystal (RIV-MAL Co) compared with that of RIV and MAL at all compaction pressures. As shown in the compactibility plot, MAL compacts exhibited reasonably good BS at comparatively larger porosity values as compared to the porosity values of RIV and RIV-MAL Co. The inability of MAL to undergo volume reduction (i.e., lower compressibility) may be a cause of its larger porosity at comparable compaction pressures.

Tabletability is defined as "the capacity of the powder material to be transformed into a tablet of specified tensile strength under the effect of applied compaction pressure" [21,42]. The overall tabletability of a material is governed by both BA and BS. The tensile strength (σ) of all the solids increased with increasing compaction pressure from 37 to 222 MPa (Figure 4c). The tensile strength value of the compacts at higher compaction pressure (222 MPa) was comparatively lowest for MAL (σ = 1.9 MPa), intermediate (slightly higher) for RIV (σ = 2.1 MPa), and the highest for RIV-MAL Co (σ = 3.2 MPa). The tabletability order based on the tensile strength at the higher compaction pressure follows the order MAL < RIV < RIV-MAL Co. The required tensile strength for a tablet to withstand the stresses encountered during its handling and transport is 2.0 MPa, and it should be attained by compressing a material at compaction pressure $\leq$ 200 MPa. Interestingly, the tabletability plot for RIV-MAL Co indicated that the cocrystal could produce a tablet of sufficient tensile strength (2.3 MPa) at a relatively lower compaction pressure of 37 MPa (please refer Figure 4c). The RIV compacts demonstrated a tensile strength of 2.1 MPa at a compaction pressure of 222 MPa, while the desired tensile strength of 2 MPa could not be achieved for MAL even at a compaction pressure as high as 222 MPa. Hence, the cocrystallization led to significant improvement in the mechanical properties, despite the fact that both parent components (RIV and MAL) possessed relatively poor mechanical properties. Hence, it is interesting to unfold the supramolecular and crystallographic basis of the compaction behavior of these three solids.

4.3.1. Heckel Analysis

The Heckel model provides method for transforming force and displacement signals to a more simpler linear relationship for materials undergoing the compaction process [43]. This increases the popularity of model for studying relationship between relative density and applied compaction pressure. [19,26,44]. The basis for the Heckel equation is that the densification of the bulk powder under applied compaction pressure follows first-order kinetics [44].

The "in-die" measurements do not account for elastic recovery of material, eventually affecting data interpretation and accuracy. Therefore, in the present study, densification of RIV, MAL, and RIV-MAL Co as a function of applied compaction pressure was evaluated by Heckel analysis using "out-of-die" measurement. The "out-of-die" measurement is more reliable method to study bulk deformation of a pharmaceutical material because the material is allowed to undergo elastic deformation prior to density analysis.

The linear region of the Heckel plot provides an important property of the material, that is mean yield pressure (Py). In the Heckel plots, correlation coefficients of R^2 > 0.98 were achieved in the case of RIV and RIV-MAL Co, while R^2 > 0.91 was achieved in the case of MAL. Among these solids, RIV-MAL Co showed the highest densification and the lowest Py (Figure 5). The Py of MAL (323 MPa) is very high as compared with RIV (133 MPa) and RIV-MAL Co (83 MPa). The lower mean yield pressure (Py) value obtained for RIV-MAL Co indicates its excellent plastic deformation under the applied compaction pressure. In the same way, RIV can also undergo plastic deformation at the normally utilized compaction pressure of 133 MPa. In contrast, MAL could not undergo plastic deformation at a compaction pressure as high as 323 MPa and also exhibited severe chipping and laminations at compaction pressure > 222 MPa. This behavior demonstrates poor plasticity of MAL during bulk deformation.

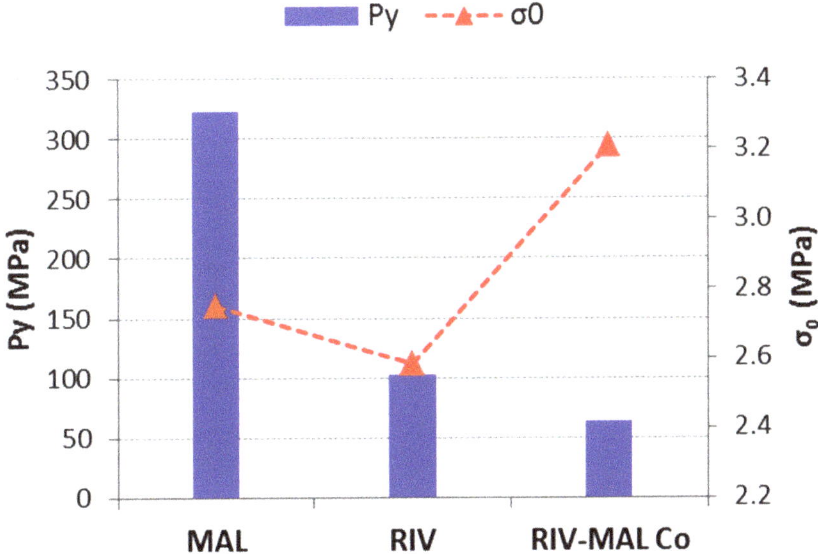

Figure 5. A plot of mean yield pressure (Py) and tensile strength at zero porosity (σ_0) for RIV, MAL, and RIV-MAL Co.

4.3.2. Ryshkewitch-Duckworth Analysis

The compactibility of the material is described by Ryshkewitch-Duckworth Equation (Equation (6) [45]). Tensile strength at zero porosity (σ_0) can be determined using this mathematical model. Since the tensile strength of a material is normalized by the bonding area at zero porosity; σ_0 represents interparticulate bonding strength (BS) of the material undergoing compaction. The effect of bonding area was minimal as the similar particle shape and PSD were used for CTC profiling.

$$\sigma = \sigma_0\, e^{-a\varepsilon} \tag{6}$$

where σ is tensile strength, α is a constant and ε is porosity. In its logarithmic form, a linear relationship between porosity and the log of the tensile strength was obtained. The value of σ_0 was 2.75, 2.58, and 3.21 MPa for MAL, RIV, and RIV-MAL Co, respectively. The higher σ_0 value indicates a greater BS for the cocrystal over the API and coformer. At the same time, the BS of MAL was higher than that of RIV. Thus, Ryshkewitch analysis supports the findings of the compactibility plot and confirmed the greater BS of the cocrystal.

The higher tabletability of RIV-MAL Co was the outcome of both higher BA and BS, as demonstrated by its low yield pressure (Py 83 MPa) and high σ_0 (3.21 MPa) (Figure 5). Comparable tabletability profiles were observed for RIV and MAL, despite the low plasticity of MAL. The tabletability of RIV was predominantly contributed by the BA, as indicated by the low yield pressure (Py of 133 MPa). BS predominantly governs the tabletability of MAL, as supported by the observed σ_0. The high Py of MAL can be correlated to its brittle nature. Based on the compactibility plots and relative tensile strength at zero porosity (σ_0), the order of BS in the three solids can be ranked as RIV < MAL < RIV-MAL Co. The Ryshkewitch–Duckworth analysis confirmed the higher BS for MAL ($\sigma_0 = 2.75$ MPa) than RIV ($\sigma_0 = 2.58$ MPa), while the highest BS was observed for RIV-MAL Co ($\sigma_0 = 3.21$ MPa). In other words, the higher work of adhesion was observed at significantly lower compaction pressure in RIV-MAL Co as compared to RIV and MAL.

4.4. Particle Level Deformation: Quantifying Crystal Deformation by Nanoindentation

Crystals with smooth surfaces were subjected to nanoindentation to decipher the particle-level deformation behaviour. The direction of the applied stress was perpendicular to the slip plane predicted based on the visualization and attachment energy calculations (applicable for RIV and RIV-MAL Co). Nanoindentation parameters, i.e., elastic modulus (E), mechanical hardness H, and 1/E values for crystal samples, are presented in Table 2.

Table 2. Elastic modulus (E), mechanical hardness H, and 1/E values for crystal samples.

Sample	H (GPa) [a]	E (GPa) [a]	1/E
RIV	0.20 (0.02)	3.41 (0.24)	0.293
MAL	0.71 (0.08)	17.91 (2.35)	0.056
RIV-MAL Co	0.51 (0.04)	17.58 (0.42)	0.057

[a] Average values are presented, while standard deviations are shown in parentheses (n = $\geq$10).

Indentation hardness (mechanical hardness) H denotes the resistance offered by the material to plastic deformation [29]. A low value of H is indicative of lower resistance offered by the material to undergo irreversible (plastic) deformation. However, organic molecular solids show initial elastic deformation followed by plastic deformation. Before plastic deformation takes place, the elastic limit has to be exceeded by the applied stress.

E is a measure of the resistance to elastic deformation and is a function of the energy of the interaction between molecules and their distances of separation [29]. The 1/E (compliance) can be correlated with elastic recovery (ER). A high ER indicates dominance of elastic deformation, which adversely contributes to plasticity (BA) and thus to the tabletability of the material.[46]

The lower the value of E, the larger is the 1/E and hence higher will be the elastic recovery. Amongst the three molecular solids, RIV possesses the lowest E (03.41 GPa), indicating the highest elastic recovery. The high elastic recovery of RIV crystals was also verified by the higher (h_{max}-h_p) value (178 GPa) (Figure 6), which is the elastic recovery determined from the (*p-h*) loading–unloading curve. The value of (h_{max}-h_p) was significantly higher for RIV (178 GPa) as compared to MAL (74 GPa) and RIV-MAL (75 GPa). The high E value (i.e., low 1/E) and low (h_{max}-h_p) for MAL and RIV-MAL Co were also indicative of lower ER of the coformer and cocrystal as compared to the API.

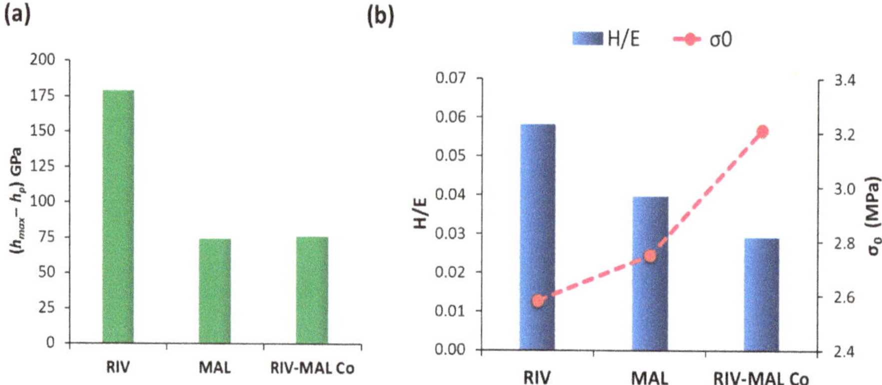

Figure 6. Nanoindentation parameters for crystal deformation representing (**a**) The (h_{max}-h_p) values depicting elastic recovery for crystals of RIV, MAL, and RIV-MAL Co; (**b**) Inverse correlation of H/E ratio (primary axis) and interparticulate bonding strength, σ_0 (secondary axis).

The high H of MAL (0.71 GPa) depicts a higher intermolecular interaction between MAL molecules, which can be directly correlated to the crystallographic features of MAL. Careful evaluation of the crystal structure of MAL showed hydrogen bonding interactions along all three axes (3D H-bonding), which provided the greater hardness to MAL (this is thoroughly discussed in Section 4.5.3). The comparatively higher hardness may hinder plastic deformation of MAL crystals when subjected to bulk compaction. RIV-MAL Co showed an "intermediate" value of H when compared with the H values of MAL and RIV (Table 2). At the same time, the cocrystal exhibited low elastic recovery as evidenced by the low values of 1/E and (h_{max}-h_p) (Figure 6). Thus, the crystals of cocrystal possess a dominance of plasticity over elasticity.

Wendy and Hewitt in 1989 found that the H/E ratio can be used to predict bulk deformation (compaction) behaviour of materials based on particle-level deformation study using the microindentation technique [31]. In this study, acetaminophen possessed the largest ratio of H/E and exhibited the poorest compaction, i.e., tablets capped and delaminated extensively during decompression and ejection from the die. Adipic acid compacts with a relatively large H/E ratio also underwent delamination during wear testing. Conversely, the materials with a lower H/E ratio could form tablets free from the above defects. This means Hewitt"s work only "qualitatively" correlated H/E ratio with compaction behaviour because the compaction behaviour of the materials was qualitatively described as "good" (for materials whose compacts were free from of capping and lamination) or "poor" (for materials whose compacts showed capping or lamination). Interestingly, the present work provided a correlation of H/E ratio with a "quantitative" bulk deformation parameter, i.e., interparticulate bonding strength at zero porosity (σ_0). In this work, the H/E was found to inversely correlate to σ_0 (Figure 6b). The cocrystal had the lowest H/E (0.029) and exhibited the highest σ_0 (3.21 MPa), while RIV had the highest H/E (0.58) and showed the lowest σ_0 (2.58 MPa). The value of H/E ratio was "intermediate" for MAL (0.040) and hence MAL exhibited the "intermediate" σ_0 (2.75 MPa) among the three molecular solids.

4.5. Identification of Crystallographic Features

Crystallographic features such as slip planes, topology of slip planes, crystallographic density, molecular packing and nature of intermolecular interactions predominantly govern the deformation behavior at both crystal as well as bulk level. The crystallographic factors influencing the bulk deformation behavior can be divided into two categories—(1) those contributing to increasing BA (plasticity) include—slip planes, topology and numbers of slip planes; (2) those contributing to BS include—the strength of interactions along the weakest crystallographic planes and true density or crystallographic true density. It is well reported that the presence of active slip planes is responsible for plastic deformation (increasing BA) of organic molecular solids including pharmaceutical solids [13,47,48]. Slip planes are defined as *"crystallographic planes in the crystal structure which contain the weakest interaction between the adjacent planes and are accounted by the highest molecular density and largest d-spacing, as compared to the other planes in that crystal"* [23,47,49].

The slip plane identification based on E_{att} calculation assumes that a plane with the least absolute attachment energy would have the weakest interaction between the adjacent planes and could slip (glide) over one another more easily than other planes in the crystal. The crystal morphology predictions by the "growth morphology" method assumes that the planes with lower attachment energies will grow at a slower rate and hence will be manifested as the larger faces of the crystal habit and vice versa. Typically, the slip plane in a crystal structure is likely to be the one with the least E_{att} and may manifest as the "largest" face within the crystal habit. The largest surface face (facet) has the highest surface area contribution to the crystal habit, and hence can be experimentally identified by either simple microscopic analysis or more accurately by face indexation analysis. Further, nanoindentation of the crystal provides experimental evidence of the presence or absence of slip planes when the stress is applied normal to the most predominant facet of a crystal (i.e., probable slip plane).

A combination of visualization and attachment energy calculations (E_{att}) could provide a more accurate prediction of the slip system compared to either individual method. Therefore, both visualization and E_{att} calculations methods were employed for reliable identification of slip systems (Table 3). E_{att} values obtained from COMPASS II force field method are presented in Table 3. The results of E_{att} calculations using "Dreiding force field method" are provided in the Supplementary Materials file. The presence and absence of slip planes in the molecular solids was experimentally confirmed by the nanoindentation study and the face indexation data previously reported by the authors [10].

Table 3. Identification of slip plane using visualization, E_{att} calculation and nanoindentation studies.

Materials	CCDC Code	Slip Planes Identification by		Nanoindentation (Is Slip System Present?)
		Visualization	Attachment Energy (hkl), E_{att} in kcal/mol $	
RIV-MAL Co	1854618	(0 0 1)	(0 0 1), −28.4	present
RIV	1854617	(0 1 $\bar{1}$) (1 0 $\bar{2}$)	(0 1 1), −44.2 (0 0 1), −39.9 (1 0 $\bar{2}$), −83.3	present
MAL	1209218	absent	(1 0 0), −13.3	absent

$ E_{att} values reported here are from COMPASS II force field method.

4.5.1. Crystallographic Features of RIV-MAL Co

The visualization and E_{att} methods predicted the slip along (0 0 1) plane in the RIV-MAL Co crystal structure (Figure 7 and Table 3). The visualization method showed the (0 0 1) plan to be a slip plane with a flat-layered topology (Table 4). The nature of hydrogen bonding is one-dimensional (1D) across each adjacent layer of the slip planes, with the (0 0 1) plane exhibiting the largest inter-planer distance (d-spacing) of 6.2127 Å when determined from simulated XRPD data. In addition to intramolecular H-bonding between RIV molecules, intermolecular H-bonding between RIV and MAL molecules was also observed in the crystal structure of RIV-MAL Co.

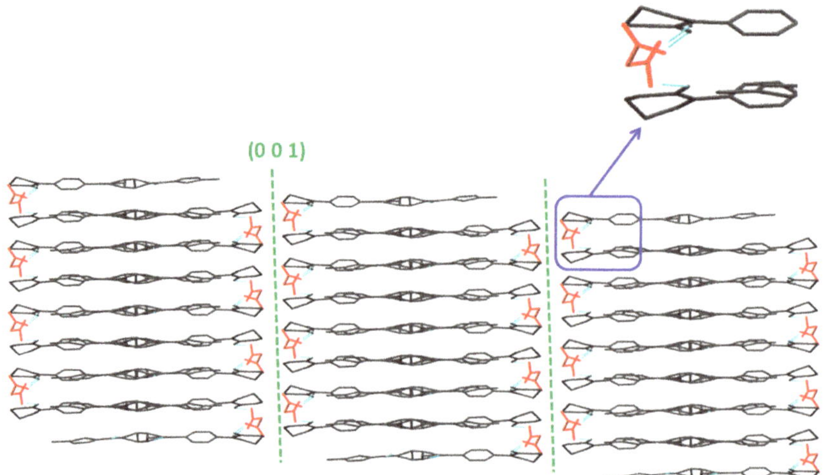

Figure 7. Crystal structure of RIV-MAL Co revealing the presence of flat layers topology slip plane along the (0 0 1) plane. MAL molecules are shown as "red", RIV in "black" and hydrogen bonding interaction between RIV and MAL as "cyan". The "insert" shows increased intermolecular interactions along the weakest crystallographic plane due to incorporation of MAL in the crystal lattice.

Table 4. Comparative assessment of molecular and supramolecular attributes of the coformer, API and cocrystal.

Materials	H-Bonding Dimensionality	No. of Slip Planes	d-Spacing (Å) and Slip Plane	Surface Topology
RIV-MAL Co	1D	1	6.2127, (0 0 1) *	Flat layers
RIV	1D	2	4.4850, (0 1 1)	Flat layers
			3.8752, (1 0 $\bar{2}$)	Corrugated/Zigzag layers
MAL	3D	0	-	Network of hydrogen bonds

* (0 0 1) plane was identified as (0 0 4) in the simulated PXRD scan of RIV.

The growth morphology model showed the (0 0 1) plane having the lowest E_{att}, indicating that the plane along this direction may have the weakest intermolecular interactions, hence it can be a probable slip system. Using the face indexing analysis, the facet corresponding to (0 0 1) plane was identified as morphologically the largest crystal facet of RIV-MAL Co, with a relative surface contribution of 57.9% to the crystal habit [10]. These results provide experimental evidence that supports the findings of visualization and E_{att} methods. Further, the "pop-ins" observed in the nanoindentation (*p-h*) loading curve of RIV-MAL Co crystal provided additional evidence for the presence of active slip plane system in the crystal structure (Figure 8).

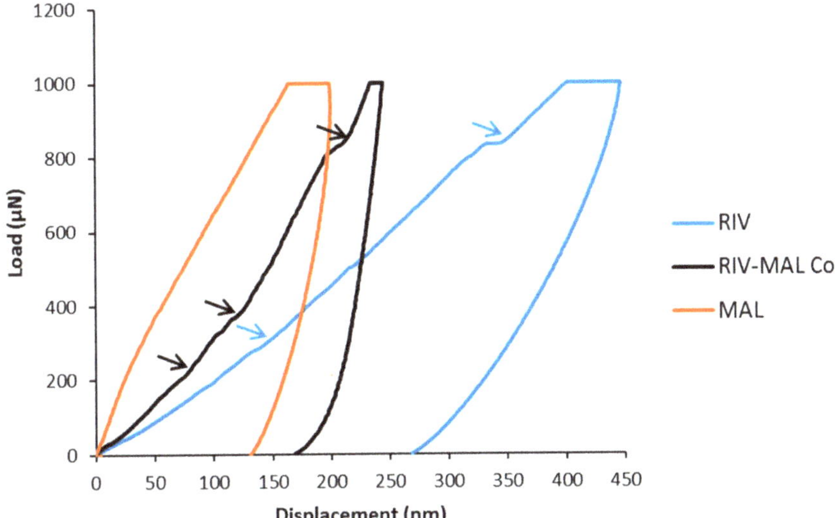

Figure 8. The load-displacement (*p-h*) curve for RIV, MAL, and RIV-MAL Co. The pop-ins observed in crystal samples of RIV and RIV-MAL Co are shown by pointing "blue" and "black", arrows, respectively.

Pop-ins (burst in) is a result of discontinuities (a sudden increase of displacement at same load) in the load-displacement (*p-h*) curve and indicates plastic deformation mediated by the active slip planes in organic molecular solids. When indention stress is applied perpendicular to the slip plane (i.e., facet corresponding to slip plane), pop-ins can be observed due to the increased plasticity presented by the slip planes that are encountered during the penetration of nanoindenter tip. Conversely, a smooth curve without pop-in is indicative of the absence of a slip-plane system in a given crystal [29].

4.5.2. Crystallographic Features of RIV

Visualization of the RIV crystal structure displayed the presence of (0 1 1) and (1 0 $\bar{2}$) as the primary and secondary slip planes, respectively (Figure 9). The plane (0 1 1) is termed as the "primary slip plane" as it has comparatively larger d-spacing (4.4850 Å) than the (1 0 $\bar{2}$) plane (3.8752 Å), which is considered as the "secondary slip-plane" (Table 4). As shown in Figure 9, the surface topology of (0 1 1) is a flat layer, while that of (1 0 $\bar{2}$) is corrugated or zigzag layers. Similar to RIV-MAL Co, the RIV crystal structure possesses intermolecular 1D H-bonding between RIV molecules.

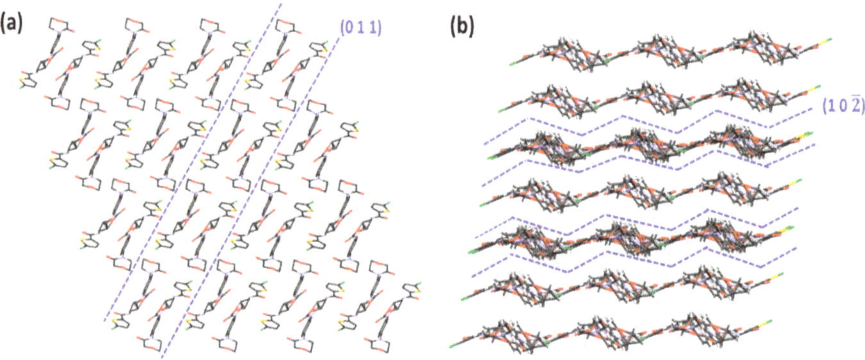

Figure 9. Crystal structure evaluation of RIV indicating the presence flat layers of RIV molecules with 1D hydrogen bonding along (0 1 1) plane (**a**), and corrugated layers along (1 0 $\bar{2}$) plane (**b**).

On the E_{att} calculations, the lower attachment energies (close to the least E_{att}) were observed for the two planes, i.e., (0 1 1) with E_{att} = −44.2 kcal/mol, and (0 0 1) with E_{att} = −39.9 kcal/mol (Table 3). This indicates the possibility of two primary active slip planes. However, when (0 0 1) plane is visualized using Mercury software, it shows strong intermolecular interactions ruling out the possibility of (0 0 1) being a slip plane. Moreover, face indexing analysis has previously reported that facet corresponding to (0 1 1) plane as morphologically the largest crystal facet with a relative surface contribution of 34.8% to the crystal habit of RIV [12]. This experimental evidence supports the presence of (0 1 1) as the primary slip plane. The E_{att} of (1 0 $\bar{2}$) plane was observed to be -83.3 kcal/mol. The presence of active slip plane in the crystal structure of RIV was further supported by the observed pop-ins in the nanoindentation (*p-h*) loading curve (Figure 8).

4.5.3. Crystallographic Features of MAL

The visualization of the crystal structure of MAL revealed the presence of interlocked 3D hydrogen bonding between MAL molecules (resembling isotropic interactions) (Figure 10). The possibility of slip (glide) is not likely in the MAL crystal structure owing to the presence of strong intermolecular interactions across all three crystallographic axes (x, y, and z). The lowest attachment energy was observed for (1 0 0) plane in the MAL crystal structure (E_{att} = −13.3 kcal/mol) (Table 3). The slip system prediction using attachment energy calculations assumes any plane with the least E_{att} as a slip plane [47,48,50]. However, this assumption may provide erroneous predictions as it does not consider the role of the strong hydrogen bonding interactions commonly present in a crystal structure with 3D interlocked molecules. Some planes in the 3D interlocked structure would obviously have lower interactions energies than other planes, hence one may consider the least attachment energy plane as the "slip plane" based on the calculated E_{att}. Therefore, prediction of slip planes in a 3D interlocked crystal structure only based on Eatt calculations can be erroneous. For 3D interlocked crystal structure, one should rely on either visualization method or other experimental techniques such as nanoindentation. The visualization method indicated the absence of a slip plane in the crystal structure of MAL. Additionally, the smooth (*p-h*) loading curve (without pop-ins) verified the absence of slip planes in the MAL crystal structure (Figure 8).

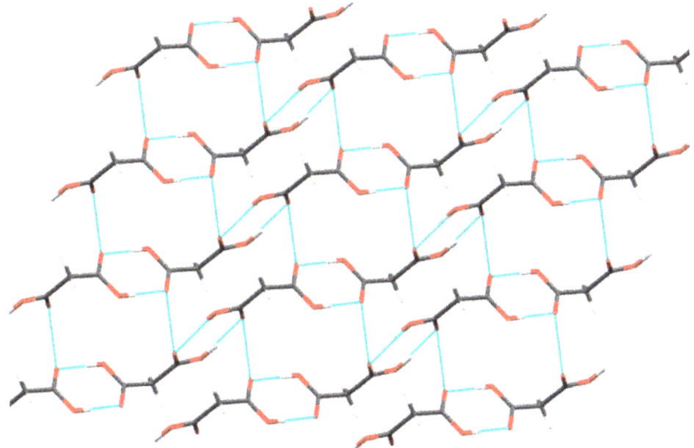

Figure 10. Crystal structure of MAL illustrating 3D hydrogen bonding network with strong intermolecular interactions.

4.6. Decoding the Basis of Bulk Deformation Behavior: Impact of Crystallographic and Supramolecular Features

4.6.1. Impact of Crystallographic Features on Plasticity

Based on the compressibility plots and mean yield pressures, the order of plasticity (i.e. BA) for the three solids is MAL < RIV < RIV-MAL Co. Thus, the observed plastic deformation can be qualitatively written as "good" for RIV-MAL Co (Py of 83 MPa), "average" for RIV (Py of 133 MPa) and "poor" for MAL (Py of 323 MPa).

The greater plasticity of the cocrystal can be ascribed to the presence of the (0 0 1) active slip plane, which facilitates the facile slip in a direction perpendicular to the applied compaction pressure. Moreover, the higher inter-planar distance (d-spacing, 6.217 Å) in the slip plane of the cocrystal crystal structure as compared to the slip plane in RIV crystal structure (d-spacing, 4.4850 Å) could help in better slippage of the planes (higher plasticity) in the cocrystal system.

The crystal structure evaluation of RIV unveiled the presence of two slip planes, i.e., (0 0 1) and (1 0 $\bar{2}$) and one-dimensional hydrogen bonding (Table 4). It was expected that RIV would have excellent plasticity owing to the presence of the multiple slip systems. However, RIV was less plastic than RIV-MAL Co. This behavior can be explained by the presence of corrugated or zigzag topology of the secondary slip plane (1 0 $\bar{2}$).

The corrugated topology has an adverse impact on plastic deformation due to possible intertwisting (interlocking) of these planes during gliding over other nearest plane(s). The unfavorable impact of corrugated slip planes on plastic deformation of materials was also reported in crystal systems like paracetamol Form I, sulfathiazole Form IV and sucrose [49]. The presence of corrugated slip planes may facilitate a new hydrogen bond formation during slippage (gliding) process. On one hand, the intermolecular interaction is weaker due to separation of layers, while on the other hand, the intertwisting (interlocking) brings two layers close enough for formation of new hydrogen bonds. The likelihood of interlocking depends upon the degree of interlayer separations [51]. The inter-planar distance, i.e., d-spacing, can be a good indicator of interlayer separation between two planes [49]. A lower value of d-spacing indicates lesser interlayer separation and vice versa. In the case of RIV, a low value of d-spacing (3.8752 Å) for (1 0 $\bar{2}$) plane signifies less interlayer separation and increased chances of interdigitation. The plastic deformation due to the primary slip plane (0 1 1) could be adversely affected by the resistance offered by the corrugated slip planes (10-2). Hence, the overall plasticity of RIV is governed by the presence of primary (flat layers topology slip) and secondary (corrugated topology slip) planes. Thus, RIV is a unique crystal system wherein both flat layers and corrugated slip planes are observed in the same system, and the BA is governed by both slip planes. Moreover, the higher elastic recovery of RIV crystals (demonstrated by nanoindentation) could have reduced the available BA during bulk deformation.

Crystal structure evaluation of MAL was carried out to investigate the cause of its profoundly lower plasticity. MAL molecules form multiple hydrogen bonds across all three crystallographic axes forming a 3D network of hydrogen bonds (Figure 10). The lower plastic deformation (Py of 323 MPa) could be a result of the absence of slip plane (Table 4) and the rigid crystal structure of MAL. The 3D hydrogen bonding network of MAL offers a closer crystal packing which confers a dense crystalline structure and the same is evidenced by its higher crystal packing density (crystallographic true density is 1.621 g/cm^2). Closer packing of the molecules along with strong intermolecular interaction (3D hydrogen bonding network) presented a rigid structure, which resists deformation under compaction pressure, resulting in the poor compressibility (poor plastic flow) of MAL. Therefore, MAL exhibited lower densification and higher yield pressure at a given pressure. The MAL possessed a lower plasticity and this material could be brittle in nature based on its high mean yield pressure (Py).

4.6.2. Impact of Crystallographic Features on BS

The measured BS in terms of σ_0 was found to be higher in MAL (σ_0 = 2.75 MPa) than RIV (σ_0 = 2.58 MPa), while the highest BS was observed for RIV-MAL Co (σ_0 = 3.21 MPa). The crystal systems devoid of slip planes showed a direct relationship between true density and compactibility (BS) and the inverse relationship between true density and compressibility (BA). This relationship holds true for the series of crystal systems studied in our lab [19,25,26] and is largely applicable to other systems [20,52]. Closely packed crystal systems have a higher crystallographic density (true density) and exhibit higher intermolecular interactions as compared to the crystal systems with an open structure and lower true density. Therefore, higher true density enables greater intermolecular contacts during compaction, resulting in higher BS. At the same time, the systems having higher true density resist densification under the applied compaction pressure and hence exhibit higher yield pressure and poor compressibility (low BA).

MAL crystal structure is devoid of slip planes and contains higher intermolecular interactions across all three axes (3D interlocked) that give rise to its higher true density of 1.621 g/cm^3. The reasonably good BS and poor compressibility (poor plasticity) of MAL were thus attributed to its higher true density. Many reported studies have considered only slip plane or plasticity (increasing BA) as key to govern tabletability and have neglected the role of BS. It is necessary to understand the role of BS in governing tabletability of molecular solids, and also considering "true density" as an important parameter, as it has a direct correlation with BS.

In crystal systems with active slip planes, BS is expected to be directly proportional to the strength of intermolecular interactions across the lowest energy slip plane in the corresponding crystal form [24,31]. In the crystal structure of RIV-MAL Co, MAL molecules form additional intermolecular hydrogen bonding interactions between the carbonyl (C=O) group of morpholinone in RIV (H-bond acceptor) and hydroxyl (-OH) in MAL (H-bond donor) [10]. Thus, cocrystallization of RIV with MAL led to the increased extent of intermolecular interactions in the crystal structure of RIV-MAL Co along the lowest energy slip plane (0 0 1). Thus, the significantly higher BS (σ_0 = 3.21 MPa) of the cocrystal over RIV (σ_0 = 2.58 MPa) can be attributed to the increased strength of intermolecular interaction. A summary of the key findings of the present work is pictorially captured in Figure 11.

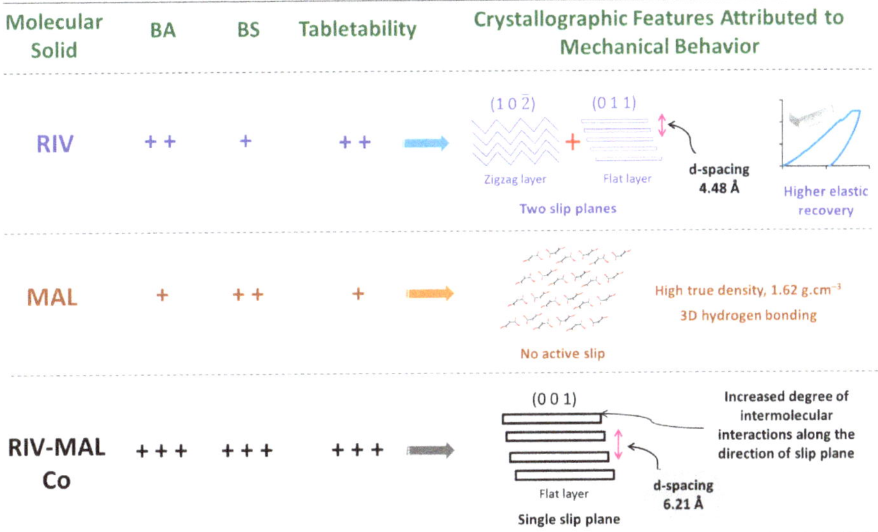

Figure 11. Pictorial depicting a summary of the key findings of present work. The observed values of interparticulate bonding strength (BS), bonding area (BA), and tabletability are expressed in "+" symbol. For the purpose of comparison, "+ + +" indicate "high"; "+" indicate "low"; and "+ +" represents "intermediate" value.

5. Conclusions

Cocrystallization led to altered crystallographic (supramolecular) features as compared to the parent components owing to the creation of a new multi-component crystal phase, which resulted to improved tabletability. This work assessed particle (crystal) and bulk level deformation behaviour of the molecular solids containing no active slip plane (i.e., MAL), a single active slip plane (i.e., RIV-MAL Co) and two slip planes of different surface topologies (RIV).

The BA in RIV was attributed to different surface topologies (flat and corrugated) of two active slip plane systems and higher elastic recovery of RIV crystals. The higher true density and the higher degree of intermolecular interactions due to the 3D interlocked structure offered reasonably good BS to MAL compacts. Concurrently, the strong intermolecular interactions resisted the densification under applied compaction pressure and hence resulted in a decrease in BA. The API and coformer displayed poor particle and bulk deformation; however, RIV-MAL compacts exhibited higher BA and greater BS. The cocrystal with highest BA and BS demonstrated significantly highest tabletability amongst the three samples. The increasing BA was attributed to the presence of flat-layered slip plane and the higher inter-planar distance, while the higher BS was ascribed to the increased degree of intermolecular interactions. During tabletability evaluation of molecular organic solids, plasticity signifies the role of the only BA. The present work stresses the necessity of understanding the role of BS as well.

The particle level deformation parameter H/E was found to inversely correlate with bulk level deformation parameter σ_0, which has been commonly used to indicate BS. We suggest the utility of this correlation for estimation of bulk deformation behaviour based on the crystal deformation behaviour studied using nanoindentation experimentation. The predictability of this relationship needs to be further verified by studying more organic molecular solids.

Supplementary Materials: The following are available online at http://www.mdpi.com/1999-4923/12/6/546/s1. Table S1. Slip plane identification by different methods. Figure S1. Heckel plot fitted for estimating Py value of for MAL. Figure S2. Heckel plot fitted for estimating Py value of RIV. Figure S3. Heckel plot fitted for estimating Py value of RIV-MAL Co.

Author Contributions: Conceptualization, and Methodology, D.P.K.; Formal Analysis, D.P.K. and V.P.; Investigation, D.P.K., V.P., A.K. and N.K.; Project administration and Supervision, A.K.B.; Writing—original draft, D.P.K. and A.K.B. All authors have read and agreed to the published version of the manuscript.

Funding: This research received no external funding.

Acknowledgments: The authors acknowledge the technical support from Ram Naresh Yadav of IIT, Ropar (India) for nanoindentation experimentation and Jayprakash A. Yadav for CTC profiling. Kailas S. Khomane is deeply acknowledged for the insightful discussion. We are also thankful to MSN Laboratories, India, for providing gift sample of rivaroxaban API. Further, Dnyaneshwar acknowledges the Department of Pharmaceuticals, Ministry of Chemicals and Fertilizers, Government of India for Research Fellowship.

Conflicts of Interest: The authors declare no conflict of interest.

References

1. Shelley, C.F. Chapter 12. Oral delivery of immediate release dosage form. In *Remington Education: Pharmaceutics*; Pharmaceutical Press: London, UK, 2014; pp. 287–313.
2. Schultheiss, N.; Newman, A. Pharmaceutical cocrystals and their physicochemical properties. *Cryst. Growth Des.* **2009**, *9*, 2950–2967. [CrossRef]
3. Desiraju, G.R.; Vittal, J.J.; Ramanan, A. *Crystal engineering: A textbook*; World Scientific Publishing Co.: Toh Tuck, Singapore, 2011.
4. Bolla, G.; Nangia, A. Pharmaceutical cocrystals: Walking the talk. *Chem. Commun.* **2016**, *52*, 8342–8360. [CrossRef] [PubMed]
5. Kale, D.P.; Zode, S.S.; Bansal, A.K. Challenges in translational development of pharmaceutical cocrystals. *J. Pharm. Sci.* **2017**, *106*, 457–470. [CrossRef] [PubMed]
6. Good, D.J.; Rodríguez-Hornedo, N.r. Solubility advantage of pharmaceutical cocrystals. *Cryst. Growth Des.* **2009**, *9*, 2252–2264. [CrossRef]
7. Yoshimura, M.; Miyake, M.; Kawato, T.; Bando, M.; Toda, M.; Kato, Y.; Fukami, T.; Ozeki, T. Impact of the dissolution profile of the cilostazol cocrystal with supersaturation on the oral bioavailability. *Cryst. Growth Des.* **2017**, *17*, 550–557. [CrossRef]
8. Jasani, M.S.; Kale, D.P.; Singh, I.P.; Bansal, A.K. Influence of drug–polymer interactions on dissolution of thermodynamically highly unstable cocrystal. *Mol. Pharm.* **2018**, *16*, 151–164. [CrossRef]
9. Trask, A.V.; Motherwell, W.D.S.; Jones, W. Physical stability enhancement of theophylline via cocrystallization. *Int. J. Pharm.* **2006**, *320*, 114–123. [CrossRef]
10. Kale, D.P.; Ugale, B.; Nagaraja, C.; Dubey, G.; Bharatam, P.V.; Bansal, A.K. Molecular basis of water sorption behavior of rivaroxaban-malonic acid cocrystal. *Mol. Pharm.* **2019**, *16*, 2980–2991. [CrossRef]
11. Shinozaki, T.; Ono, M.; Higashi, K.; Moribe, K. A novel drug-drug cocrystal of levofloxacin and metacetamol: Reduced hygroscopicity and improved photostability of levofloxacin. *J. Pharm. Sci.* **2019**, *108*, 2383–2390. [CrossRef]
12. Yadav, J.P.; Yadav, R.N.; Uniyal, P.; Chen, H.; Wang, C.; Sun, C.C.; Kumar, N.; Bansal, A.K.; Jain, S. Molecular interpretation of mechanical behavior in four basic crystal packing of Isoniazid with homologous cocrystal formers. *Crystal Growth & Design* **2020**, *20*, 832–844.
13. Reddy, C.M.; Krishna, G.R.; Ghosh, S. Mechanical properties of molecular crystals—applications to crystal engineering. *CrystEngComm* **2010**, *12*, 2296–2314. [CrossRef]
14. Karki, S.; Frisic, T.; Fabian, L.; Laity, P.R.; Day, G.M.; Jones, W. Improving mechanical properties of crystalline solids by cocrystal formation: New compressible forms of paracetamol. *Adv. Mater.* **2009**, *21*, 3905–3909. [CrossRef]
15. Krishna, G.R.; Shi, L.; Bag, P.P.; Sun, C.C.; Reddy, C.M. Correlation among crystal structure, mechanical behavior, and tabletability in the co-crystals of vanillin isomers. *Cryst. Growth Des.* **2015**, *15*, 1827–1832. [CrossRef]
16. Aher, S.; Dhumal, R.; Mahadik, K.; Ketolainen, J.; Paradkar, A. Effect of cocrystallization techniques on compressional properties of caffeine/oxalic acid 2: 1 cocrystal. *Pharm. Dev. Technol.* **2013**, *18*, 55–60. [CrossRef]
17. Hiendrawan, S.; Veriansyah, B.; Widjojokusumo, E.; Soewandhi, S.N.; Wikarsa, S.; Tjandrawinata, R.R. Physicochemical and mechanical properties of paracetamol cocrystal with 5-nitroisophthalic acid. *Int. J. Pharm.* **2016**, *497*, 106–113. [CrossRef]

18. Pedersen, S.; Kristensen, H. Compaction behaviour of 4-hydroxybenzoic acid and two esters compared to their mechanical properties. *Eur. J. Pharm. Biopharm.* **1995**, *41*, 323–328.

19. Khomane, K.S.; More, P.K.; Raghavendra, G.; Bansal, A.K. Molecular understanding of the compaction behavior of indomethacin polymorphs. *Mol. Pharm.* **2013**, *10*, 631–639. [CrossRef]

20. Picker-Freyer, K.M.; Liao, X.; Zhang, G.; Wiedmann, T.S. Evaluation of the compaction of sulfathiazole polymorphs. *J. Pharm. Sci.* **2007**, *96*, 2111–2124. [CrossRef]

21. Sun, C.C. Decoding powder tabletability: Roles of particle adhesion and plasticity. *J. Adhes. Sci. Technol.* **2011**, *25*, 483–499. [CrossRef]

22. Osei-Yeboah, F.; Chang, S.-Y.; Sun, C.C. A critical examination of the phenomenon of bonding area-bonding strength interplay in powder tableting. *Pharm. Res.* **2016**, *33*, 1126–1132. [CrossRef]

23. Joiris, E.; Di Martino, P.; Berneron, C.; Guyot-Hermann, A.-M.; Guyot, J.-C. Compression behavior of orthorhombic paracetamol. *Pharm. Res.* **1998**, *15*, 1122–1130. [CrossRef] [PubMed]

24. Sun, C.C.; Hou, H. Improving mechanical properties of caffeine and methyl gallate crystals by cocrystallization. *Cryst. Growth Des.* **2008**, *8*, 1575–1579. [CrossRef]

25. Khomane, K.S.; More, P.K.; Bansal, A.K. Counterintuitive compaction behavior of clopidogrel bisulfate polymorphs. *J. Pharm. Sci.* **2012**, *101*, 2408–2416. [CrossRef] [PubMed]

26. Upadhyay, P.; Khomane, K.S.; Kumar, L.; Bansal, A.K. Relationship between crystal structure and mechanical properties of ranitidine hydrochloride polymorphs. *CrystEngComm* **2013**, *15*, 3959–3964. [CrossRef]

27. Egart, M.; Janković, B.; Srčič, S. Application of instrumented nanoindentation in preformulation studies of pharmaceutical active ingredients and excipients. *Acta Pharmaceutica* **2016**, *66*, 303–330. [CrossRef]

28. Cao, X.; Morganti, M.; Hancock, B.C.; Masterson, V.M. Correlating particle hardness with powder compaction performance. *J. Pharm. Sci.* **2010**, *99*, 4307–4316. [CrossRef]

29. Varughese, S.; Kiran, M.; Ramamurty, U.; Desiraju, G.R. Nanoindentation in crystal engineering: Quantifying mechanical properties of molecular crystals. *Angew. Chem. Int. Ed.* **2013**, *52*, 2701–2712. [CrossRef]

30. Jing, Y.; Zhang, Y.; Blendell, J.; Koslowski, M.; Carvajal, M.T. Nanoindentation method to study slip planes in molecular crystals in a systematic manner. *Cryst. Growth Des.* **2011**, *11*, 5260–5267. [CrossRef]

31. Duncan-Hewitt, W.C.; Weatherly, G.C. Evaluating the hardness, Young's modulus and fracture toughness of some pharmaceutical crystals using microindentation techniques. *J. Mater. Sci. Lett.* **1989**, *8*, 1350–1352. [CrossRef]

32. Yadav, J.P.; Yadav, R.N.; Sihota, P.; Chen, H.; Wang, C.; Sun, C.C.; Kumar, N.; Bansal, A.; Jain, S. Single-crystal plasticity defies bulk-phase mechanics in isoniazid cocrystals with analogous coformers. *Cryst. Growth Des.* **2019**, *19*, 4465–4475. [CrossRef]

33. Persson, A.-S.; Ahmed, H.; Velaga, S.; Alderborn, G. Powder compression properties of paracetamol, paracetamol hydrochloride, and paracetamol cocrystals and coformers. *J. Pharm. Sci.* **2018**, *107*, 1920–1927. [CrossRef] [PubMed]

34. Phadke, C.; Sharma, J.; Sharma, K.; Bansal, A.K. Effect of variability of physical properties of povidone K30 on crystallization and drug–polymer miscibility of celecoxib–povidone K30 amorphous solid dispersions. *Mol. Pharm.* **2019**, *16*, 4139–4148. [CrossRef] [PubMed]

35. Oliver, W.C.; Pharr, G.M. Measurement of hardness and elastic modulus by instrumented indentation: Advances in understanding and refinements to methodology. *J. Mater. Res.* **2004**, *19*, 3–20. [CrossRef]

36. Berkovitch-Yellin, Z. Toward an ab initio derivation of crystal morphology. *J. Am. Chem. Soc.* **1985**, *107*, 8239–8253. [CrossRef]

37. Docherty, R.; Clydesdale, G.; Roberts, K.; Bennema, P. Application of Bravais-Friedel-Donnay-Harker, attachment energy and Ising models to predicting and understanding the morphology of molecular crystals. *J. Phys. D: Appl. Phys.* **1991**, *24*, 89. [CrossRef]

38. Caires, F.J.; Lima, L.; Carvalho, C.; Giagio, R.; Ionashiro, M. Thermal behaviour of malonic acid, sodium malonate and its compounds with some bivalent transition metal ions. *Thermochim. Acta* **2010**, *497*, 35–40. [CrossRef]

39. Khomane, K.S.; Bansal, A.K. Weak hydrogen bonding interactions influence slip system activity and compaction behavior of pharmaceutical powders. *Journal of pharmaceutical sciences* **2013**, *102*, 4242–4245. [CrossRef]

40. Sun, C.C.; Grant, D.J. Improved tableting properties of p-hydroxybenzoic acid by water of crystallization: A molecular insight. *Pharm. Res.* **2004**, *21*, 382–386. [CrossRef]

41. Sun, C.; Himmelspach, M.W. Reduced tabletability of roller compacted granules as a result of granule size enlargement. *Journal of Parmaceutical Sciences* **2006**, *95*, 200–206. [CrossRef]

42. Khomane, K.S.; Bansal, A.K. Differential compaction behaviour of roller compacted granules of clopidogrel bisulphate polymorphs. *Int. J. Pharm.* **2014**, *472*, 288–295. [CrossRef]

43. Patel, S.; Kaushal, A.M.; Bansal, A.K. Compression physics in the formulation development of tablets. *Critical Reviews™ in Therapeutic Drug Carrier Systems* **2006**, *23*, 1–66. [CrossRef] [PubMed]

44. Heckel, R. Density-pressure relationships in powder compaction. *Trans. Metall. Soc. AIME* **1961**, *221*, 671–675.

45. Ryshkewitch, E. Compression strength of porous sintered alumina and zirconia: 9th communication to ceramography. *J. Am. Ceram. Soc.* **1953**, *36*, 65–68. [CrossRef]

46. Govedarica, B.; Ilić, I.; Šibanc, R.; Dreu, R.; Srčič, S. The use of single particle mechanical properties for predicting the compressibility of pharmaceutical materials. *Powder Technol.* **2012**, *225*, 43–51. [CrossRef]

47. Roberts, R.; Rowe, R.; York, P. The relationship between indentation hardness of organic solids and their molecular structure. *Journal of materials science* **1994**, *29*, 2289–2296. [CrossRef]

48. Bandyopadhyay, R.; Grant, D.J. Plasticity and slip system of plate-shaped crystals of L-lysine monohydrochloride dihydrate. *Pharm. Res.* **2002**, *19*, 491–496. [CrossRef]

49. Shariare, M.H.; Leusen, F.J.; de Matas, M.; York, P.; Anwar, J. Prediction of the mechanical behaviour of crystalline solids. *Pharm. Res.* **2012**, *29*, 319–331. [CrossRef]

50. Sun, C.C.; Kiang, Y.H. On the identification of slip planes in organic crystals based on attachment energy calculation. *J. Pharm. Sci.* **2008**, *97*, 3456–3461. [CrossRef] [PubMed]

51. Bryant, M.; Maloney, A.; Sykes, R. Predicting mechanical properties of crystalline materials through topological analysis. *CrystEngComm* **2018**, *20*, 2698–2704. [CrossRef]

52. Ragnarsson, G.; Sjögren, J. Compressibility and tablet properties of two polymorphs of metoprolol tartrate. *Acta Pharm. Suec.* **1984**, *21*, 321–330.

 *pharmaceutics*

Article

A Comparative Study of the Effect of Different Stabilizers on the Critical Quality Attributes of Self-Assembling Nano Co-Crystals

Bwalya A. Witika [1], Vincent J. Smith [2] and Roderick B. Walker [1,*]

[1] Division of Pharmaceutics, Faculty of Pharmacy, Rhodes University, Makhanda 6140, South Africa; bwawitss@gmail.com

[2] Department of Chemistry, Faculty of Science, Rhodes University, Makhanda, 6140 South Africa; v.smith@ru.ac.za

* Correspondence: r.b.walker@ru.ac.za

Received: 17 January 2020; Accepted: 11 February 2020; Published: 23 February 2020

Abstract: Lamivudine (3TC) and zidovudine (AZT) are antiviral agents used orally to manage HIV/AIDS infection. A pseudo one-solvent bottom-up approach was used to develop and produce nano co-crystals of 3TC and AZT. Equimolar amounts of 3TC dissolved in de-ionized water and AZT in methanol were rapidly injected into a pre-cooled vessel and sonicated at 4 °C. The resultant suspensions were characterized using a Zetasizer. The particle size, polydispersity index and Zeta potential were elucidated. Further characterization was undertaken using powder X-ray diffraction, Raman spectroscopy, Fourier transform infrared spectroscopy, differential scanning calorimetry, and energy dispersive X-ray spectroscopy scanning electron microscopy. Different surfactants were assessed for their ability to stabilize the nano co-crystals and for their ability to produce nano co-crystals with specific and desirable critical quality attributes (CQA) including particle size (PS) < 1000 nm, polydispersity index (PDI) < 0.500 and Zeta potential (ZP) < −30 mV. All surfactants produced co-crystals in the nanometer range. The PDI and PS are concentration-dependent for all nano co-crystals manufactured while only ZP was within specification when sodium dodecyl sulfate was used in the process.

Keywords: nano co-crystals; crystal engineering; polydispersity index; zeta potential; particle size; zidovudine; lamivudine; HIV/AIDS; sonochemistry

---

1. Introduction

More than 7000 people worldwide die of HIV-related causes daily. Many people are not benefiting fully from the use of orally administered antiretroviral (ARV) drugs, which provide the only effective means of halting the progression of HIV disease and AIDS [1]. Eight million of the estimated 37 million HIV-positive people should be treated, but only two million are currently receiving ARV therapy. This unmet need is expected to increase on an annual basis [1].

Crystal engineering is described as the exploitation of non-covalent interactions between molecular or ionic components for the rational design of solid-state materials [2,3]. The application of crystal engineering in pharmaceutics is usually related to understanding polymorphism and its associated properties.

Co-crystals are single-phase crystalline solids that are composed of two or more different molecules, which generally associate in a stoichiometric ratio [4]. Co-crystals can be constructed using several types of molecular interactions such as hydrogen bonds, halogen bonds, π–π stacking, van der Waal's forces, amongst others [5–8]. They are thermodynamically more stable than crystals of the pristine compounds, while for pharmaceutical applications they are highly promising for tailoring the properties of the active pharmaceutical ingredient (API) [9]. Co-crystals are known to exhibit different properties from the parent compounds including enhanced solubility, improved dissolution kinetics, improved bioavailability as well as increased phase stability when compared to amorphous forms, which tend to spontaneously crystallize on standing. Co-crystal formation does not involve or require covalent bond formation or breaking and usually requires rather mild conditions during synthesis. Solid-state synthetic methods such as neat grinding, liquid-assisted grinding, and other mechanochemical methods have recently come into prominence as reliable methods for co-crystal synthesis and because they are inherently green methods capable of producing high yields without the need for large or excessive quantities of solvent [10].

Co-formers are molecules that are selected to co-crystallize with an API and are chosen from the, "generally regarded as safe" list (GRAS) or the, "everything added to food in the United States" list (EAFUS) [11]. They include but are not limited to food additives, preservatives, pharmaceutical excipients, and other API molecules [7,9].

Finally, co-crystallization of important API molecules may lead to patents or intellectual property protection emanating from their development [12].

Despite the advantages of co-crystallization, further benefit can be derived by combining different technologies to ensure targeted drug delivery, enhanced bioavailability, flexibility in respect of administration and stealth delivery. Combining co-crystallization with nano-sizing to yield nano co-crystals presents such an opportunity. Several techniques can be used to develop co-crystals with nano-scale dimensions, many of which are derived from techniques used in nanocrystal manufacture. Nanocrystals can be manufactured using two approaches, namely: a top-down technique that uses shear forces to reduce the particle size from micrometer to nanometer dimensions [13,14] and a bottom-up approach that involves nucleation and crystal growth. The growth of individual crystals can be arrested in the nanometer range by using a suitable stabilizer [15,16].

The use of surfactants as stabilizers has previously been explored in the synthesis of nanocrystals [17–19] and nano co-crystals [20–22]. Nano co-crystals are co-crystals of nano-scale dimensions which exhibit properties that are superior to those generally associated with co-crystals and nanocrystals [20,23,24]. Stabilizers are primarily used as growth prevention agents and function by the adsorption of surfactant/polymer molecules onto nucleated nanocrystals or co-crystals, lowering the surface free energy and consequently particle reactivity [25]. Known stabilizers include surfactants such as sodium dodecyl sulfate (SDS) [26–28], Tween® [29–31], Span® [20,32], α-tocopheryl polyethylene glycol succinate 1000 (TPGS 1000) [33,34], Pluronic® [35,36] and polymers such as hydroxypropyl methylcellulose (HPMC) [19,37], pyrrolidone K30 [37] and polyvinyl pyrrolidone [19]. Different stabilizers impart different properties to the resultant nano co-crystals. For instance, TPGS 1000 is known to inhibit P-glycoprotein efflux and stealth properties to formulations in which it has been incorporated, while Tween® 80 facilitates brain targeting [31,34,38,39] and Span® is effective in reducing the size of nano co-crystals [20].

The use of a combination of techniques in the manufacture of nano co-crystals has been applied with success on a few occasions. For example, a top-down high-pressure homogenization technique (HPH) was used to produce nano co-crystals of the flavonoid, baicalein with nicotinamide. The resultant nano co-crystal exhibited a marked improvement in the rate and extent of dissolution [22]. Similarly, a bottom-up approach was used to develop myricetin-nicotinamide nano co-crystals. The nano co-crystal product also displayed an increased rate and extent for dissolution [40].

Sonochemical co-crystallization is a bottom-up process that has been successfully used for nano co-crystal synthesis [20,40–42]. One solvent systems involve dissolving all the co-crystal components in one solvent and injecting the solution into an anti-solvent while simultaneously sonicating the solution [41]. Two solvent systems involve dissolving the components of the co-crystal separately in different solvents followed by injecting each solution into the same anti-solvent at the same time. Top-down approaches such as wet media milling [40] and high-pressure homogenization [22], which are not covered in this work, have also been used with success.

Attempts have been made to produce co-crystals with nano-scale dimensions with varying success. The pharmaceutical compound caffeine and the co-former 2,4-dihydroxybenzoic acid were co-crystallized via sonochemical synthesis and stabilized with the surfactant Span® 85. The resultant co-crystal dimensions for the smallest particles were 190 × 200 nm while the largest particles had dimensions of 200 × 800 nm [20]. The presence of surfactant was found to promote nucleation and moderate crystal growth. Myricetin-nicotinamide nano co-crystals were synthesised via both bottom-up and top-down approaches using Tween® 80 as the stabiliser. The smallest particles had dimensions of 100 × 200 nm, whereas the largest particles were 200 × 800 nm [40].

Baicalein-nicotinamide (BE-NCT) nano co-crystals were successfully prepared via a top-down approach. The BE-NCT nano co-crystals were compared with BE coarse powder, BE-NCT co-crystals and BE nanocrystals, BE-NCT nano co-crystals exhibited a significantly enhanced performance both in in vitro and in vivo evaluations, suggesting that the nano co-crystals could be proposed as an advanced strategy for dissolution rate and bioavailability enhancement of poor soluble natural products such as BE [22].

Bhatt et al. synthesized a co-crystal of the ARV compounds 3TC and AZT using slow evaporation from a variety of solvents and several other methods including liquid assisted grinding [42]. The resultant co-crystal (3TC·AZT·H_2O) contains a molecule of water, one molecule of 3TC and one molecule of AZT. Each AZT molecule hydrogen bonds via N–H···O and O–H···O interactions to three different 3TC molecules as well as to two different H_2O molecules while each 3TC molecule hydrogen bonds to a single H_2O molecule via an O–H···N interaction [43].

A preliminary investigation into the possibility of using a one or two solvent approach to producing nano co-crystals of 3TC and AZT proved unsuccessful. This was probably owing to the vastly different solubilities of the API molecules. This led to the development of a pseudo one solvent approach in which both components were dissolved separately in different solvents so that each solvent serves as an anti-solvent for the other, in situ.

Nano co-crystals are considerably easy to produce but stabilizing them against continual growth after forming and stabilizer(s) selection, are critical [44]. Herein, we report the use of surfactants in combination with sonochemical methods to synthesize nanometer sized co-crystals and to investigate the impact of different surfactants on the critical quality attributes (CQA) of the resultant nano co-crystal particles. To the best of our knowledge, this is the first comparative investigation of four stabilizers, viz., Tween® 80, Span® 80, SDS and TPGS 1000 and their effect on the three CQA parameters: particle size (PS), polydispersity index (PDI) and Zeta potential (ZP) for the reported co-crystals.

2. Materials and Methods

2.1. Materials

AZT and 3TC were purchased from China Skyrun Co. Ltd. (Taizhou, China). Tween® 80, Span® 80, SDS and TPGS 1000 were purchased from Merck (Johannesburg, South Africa). HPLC-grade water was prepared by reverse osmosis using a RephiLe® Direct-Pure UP and RO water system Microsep® (Johannesburg, South Africa) fitted with a RephiDuo® H PAK de-ionization cartridge and a RephiDuo® PAK polishing cartridge. The water was filtered through a 0.22 μm PES high flux capsule filter Microsep® (Johannesburg, South Africa) prior to use. HPLC grade Honeywell Burdick and Jackson™ methanol (MeOH) was purchased from Anatech Instruments (Johannesburg, South Africa).

2.2. Methods

2.2.1. Preparation of Micro and Nano Co-Crystals Using a Pseudo One Solvent Bottom-Up Method

Micro co-crystals of AZT and 3TC have been reported and were synthesized according to previously described methods [44]. The micro co-crystals produced were used as reference material in the characterization experiments in order to elucidate the characteristics of nano co-crystal formation in this investigation. A quantity of 3TC and AZT equivalent to 2 mmol of each ARV was accurately weighed using a model AG 135 Mettler Toledo (Greifensee, Switzerland) analytical balance and dissolved in 10 mL of water and 5 mL of ethanol (EtOH) respectively. The two solutions were mixed and gently stirred at 50 °C for an hour. The solution was allowed to cool to ambient temperature (22 °C) for 48 h to allow micro co-crystals to grow.

Supersaturation studies were conducted by adding 1 mL of MeOH to the 2 mmol of AZT and similarly, 1 mL of water was added to 2 mmol of 3TC. The individual solutions were then sonicated for 5 min using a Branson® 8510E-MT ultrasonic bath (Danbury, CT, USA). Subsequently, 1 mL aliquots of solvent were added and further sonicated for 5 min until a clear solution resulted.

Nano co-crystals (NCC) of AZT and 3TC were prepared using a cold-sonochemical precipitation bottom-up technique [16,20]. The batch size was approximately 1 g, specifically 534 mg of AZT and 458 mg of 3TC amounting to the use of 2 mmol of each API. The 3TC was dissolved in 7 mL of water and AZT was dissolved in 6 mL of MeOH. The solutions were rapidly injected into a pre-cooled conical flask incubated at 4 °C ± 2 °C in an ice bath. A sonication output of 50 kHz ± 6 kHz was applied to the solution for 20 min using a Branson® 8510E-MT ultrasonic bath.

2.2.2. Particle Size Analysis

The mean PS and PDI of the NCC was determined using a Nano-ZS 90 Zetasizer (Malvern Instruments, Worcestershire, UK) with the instrument set to Photon Correlation Spectroscopy (PCS) mode. Approximately 30 µL of an aqueous dispersion of NCC was diluted with 10 mL HPLC-grade water prior to the analysis. The sample was placed into a $10 \times 10 \times 45$ mm polystyrene cell and all measurements were performed in replicate ($n = 6$) at 25 °C at a scattering angle of 90°. Analysis of PCS data was undertaken using Mie theory with real and imaginary refractive indices set at 1.456 and 0.01.

2.2.3. Zeta Potential

The ZP of the NCC was measured using a Nano-ZS 90 Zetasizer set in the Laser Doppler Anemometry (LDA) mode (replicates $n = 6$). The samples were prepared for analysis as described in Section 2.2.2 and placed into folded polystyrene capillary cells prior to measurement.

2.2.4. FTIR Spectroscopy

The IR absorption spectrum of uncoated NCC was generated using a 100 Spectrum FTIR ATR spectrophotometer (PerkinElmer, Beaconsfield, UK) and analyzed using Peak® version 4.00 spectroscopy software (Operant LLC, Burke, VA, USA). Approximately 5 mg of powder was placed onto a diamond crystal and analyzed over the wavenumber range 4000–650 cm^{-1} at a rate of 4 cm^{-1} (replicates $n = 5$) and the spectrum for the micro co-crystals was used for reference purposes.

2.2.5. Raman Spectroscopy

The Raman spectra of uncoated NCC was collected using a Bruker Ram II spectrometer (Billerica, MA, USA) and analyzed using Peak® version 4.00 spectroscopy software. Approximately 5 mg of material was placed into a stainless-steel cup and analyzed over the wavenumber range 4000–50 cm^{-1} (replicates $n = 6$) and the spectrum of the micro co-crystal was used for reference purposes.

2.2.6. Differential Scanning Calorimetry

Approximately 4 mg of ultra-filtered and dried NCC was placed into aluminum pans and sealed. The pans were then placed directly into the furnace of a DSC 6000 PerkinElmer Differential Scanning Calorimeter (Waltham, MA, USA) and the data analyzed using version 11 Pyris™ Manager Software (PerkinElmer, Waltham, MA, USA). The temperature of the DSC was monitored with a computer and a controlled heating rate of 10 K/min was used for the analysis over the temperature range 30–150 °C. Thermograms were acquired at a rate of 10 scans/cm^{-1}. All DSC analyses were conducted in triplicate

(n = 3) under a nitrogen atmosphere purged at a flow rate of 20 mL/min and the thermogram for the micro co-crystal was used for reference purposes.

2.2.7. Powder X-ray Diffraction (PXRD)

X-ray powder diffraction patterns were measured using a Bruker D8 Discover diffractometer (Billerica, MA, USA) equipped with a proportional counter, using Cu Kα radiation with a wavelength λ = 1.5405 Å and a nickel filter. All samples were placed onto a silicon wafer for the measurement of the diffraction pattern. The generator was set at 30 kV and the current to 40 mA. Replicate data (n = 3) was collected in the 2θ = 10 to 50° range at a scanning rate of 1.5 min^{-1} with a filter time-constant of 0.38 s per step and slit width of 6.0 mm. The X-ray diffraction data were treated using evaluation curve fitting (Eva) software. Baseline correction was performed on each diffraction pattern by subtracting a spline function fitted to the curved background. The diffractogram of the micro co-crystal was used for reference purposes.

2.2.8. Energy Dispersive X-ray Spectroscopy Scanning Electron Microscopy

Energy-dispersive X-ray spectroscopy EDX, often also referred to as EDS, is based on the generation of X-rays following interaction of an electron beam with sample atoms. Apart from a continuous spectrum of X-rays generated by deceleration of beam electrons due to interaction with the atoms in a sample, sharp X-ray signals are produced at wavelengths that are specific for a given element. These signals form the basis for elemental mapping by SEM-EDX [45,46].

Elemental analysis was performed using a Vega® Scanning Electron Microscope (Tuscan, Czechoslovakia Republic) fitted with an INCA PENTA FET. Approximately 1 mg of the NCC was dusted onto a graphite plate and the sample irradiated using SEM at an accelerated voltage of 20 kV (n = 3).

2.2.9. Preparation of Surfactant-Coated Nano Co-Crystals via a Pseudo One-Solvent Bottom-Up Method

Surfactant-coated nano co-crystals were prepared as described in Section 2.2.1. Design Expert® software version 8.0.71, Stat-Ease Inc. (Minneapolis, MN, USA) was used to generate experiments for a general factorial experimental design and the process factors investigated are listed in Table 1. The surfactants were added to the aqueous phase with the exception of Span® 80, which was dissolved in methanol. The solutions were rapidly injected into a pre-cooled conical flask incubated at 4 °C ± 2 °C in an ice bath. A sonication output of 50 kHz ± 6 kHz was applied to the solution for 20 min using a Branson® 8510E-MT ultrasonic bath (Danbury, CT, USA). The NCC suspension that was produced was characterized within 24 h of preparation. Characterization of the NCC included ZP, PDI, PS and FTIR, Raman spectroscopy, DSC, PXRD and EDX-SEM. The concentration of each of the surfactants was 0.5%, 1% and 2% *w/v* of the total volume used (Table 1).

Table 1. Summary of general factorial experiments.

Std. Run	Run No.	Surfactant	Concentration % *w/v*
8	1	TPGS 1000	1
2	2	SDS	0.5
11	3	Span 80	2
10	4	SDS	2
7	5	Span 80	1
5	6	Tween 80	1
9	7	Tween 80	2
4	8	TPGS 1000	0.5
12	9	TPGS 1000	2
3	10	Span 80	0.5
1	11	Tween 80	0.5
6	12	SDS	1

3. Results

3.1. Co-Crystal Synthesis

The size of the co-crystals produced was smaller than that of the starting compounds when using the bottom up approach and the size reduction yielded sub-micron crystals. The uncoated micro cocrystals have a PS of 1593 ± 148 nm, PDI of 0.751 ± 0.063 and ZP of -6.86 ± 1.04 mV. The characterization of these crystals is reported herein.

3.2. Co-Crystal Characterization

3.2.1. FTIR Spectroscopy

The FTIR spectra depicted in Figure 1 have peaks at 3530 cm^{-1} for both nano and micro co-crystals, which is characteristic peak for water in the crystal structure [47,48]. The stretching band occurring at 1634 cm^{-1} is due to the carbonyl moiety (O=C–NR$_2$) and is characteristic of 3TC. It partially overlaps with the band due to N–H bending at 1607 cm^{-1}. The band at 1648 cm^{-1} is due to the stretching vibration of the imine group (R$_2$-C=NR). Broad bands due to the stretching vibration of –NH$_2$ and –OH functional groups are observed at 3300–3500 cm^{-1} and are indicative of 3TC in the co-crystal. Characteristic bands at 2170 cm^{-1} and 1652 cm^{-1} are due to –N$_3$ and –N–H stretching vibrations and are indicative of AZT in the co-crystal.

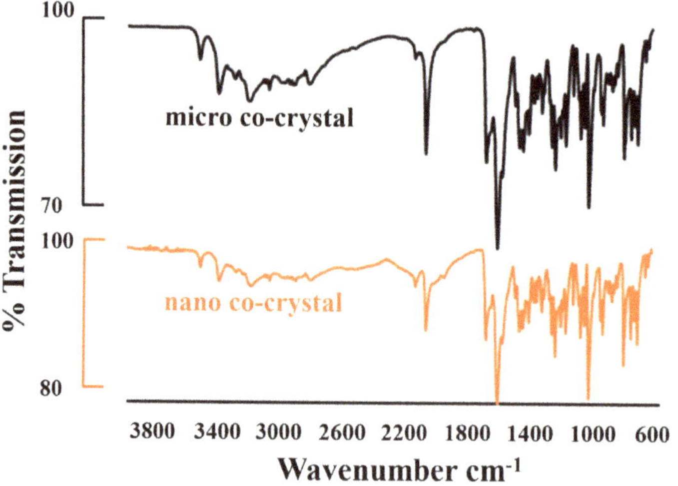

Figure 1. FTIR absorption spectra of the micro (black) and the nano co-crystal (orange).

3.2.2. Raman Spectroscopy

The Raman spectra illustrated in Figure 2 show characteristic peaks for both 3TC and AZT. In the region < 1500 cm^{-1} the Raman spectrum for 3TC exhibits several unique bands at approximately 1290 cm^{-1}, 1250 cm^{-1} and 790 cm^{-1} [49]. A carbonyl stretching mode is observed at 1650 cm^{-1} in addition to C=N stretching at 1530 cm^{-1}, both of which are confirmed in previous reports [49,50]. The most intense bands for AZT at 1650 cm^{-1} are due to the symmetric stretching vibration of the C=C bond of the pyrimidine ring. Another marker band for AZT is that corresponding to the breathing vibration of the pyrimidine ring located at 760 and at 790 cm^{-1}. The peak at 1480 cm^{-1} is due to the carbonyl C=O stretching vibration of the pyrimidine ring. The band for the N≡N stretching vibration of the azide group present in the Raman spectrum at 2100 cm^{-1} is characteristic of AZT. The signals observed are in close agreement with previously reported data [51].

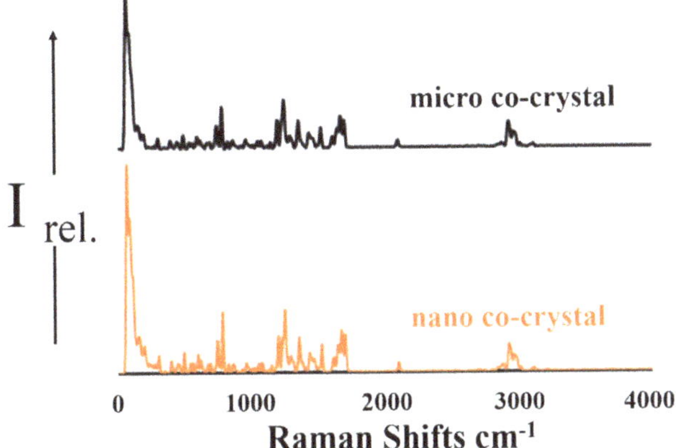

Figure 2. A stacked plot of the Raman spectra for the micro (black) and nano co-crystals (orange).

3.2.3. Differential Scanning Calorimetry

The thermograms depicted in Figure 3 show the melting endotherms of the micro and nano co-crystalline material with T_{peak} = 106.7 °C and 103.1 °C respectively. The lower melting temperature and peak broadening for the nano co-crystal is likely due to particle size reduction and is in agreement with the principles of the Van't Hoff equation [52–54]. In addition, slight peak broadening can be attributed to a loss crystallinity and partial formation of amorphous material during the formation of nanosuspensions [55,56].

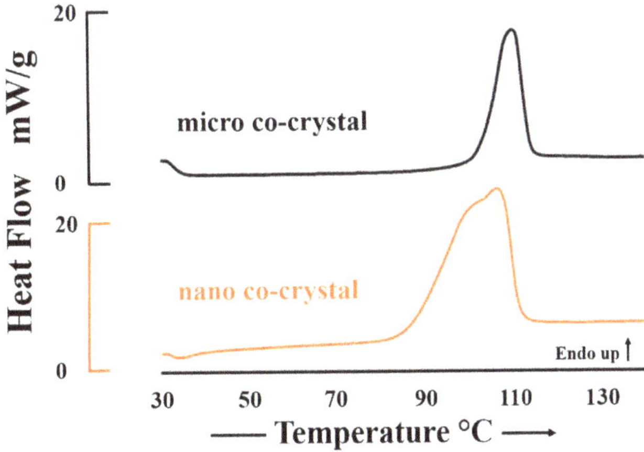

Figure 3. DSC thermograms depicting the melting endotherm for the micro (black) and nano co-crystal (orange).

3.2.4. Powder X-ray Diffraction

It is apparent from Figure 4 that there is almost a one-to-one agreement between the diffractograms for the micro and nano co-crystal of 3TC·AZT·H$_2$O. The only differences between the diffractograms relates to differences in the fine structure of the profiles. These differences may be attributed to particle size reduction of the NCC as the smaller the particle size results in broader peaks, which equates to less detail in the fine structure. Differences in the relative intensities of the peaks can be attributed to minor preferred orientation effects [57,58].

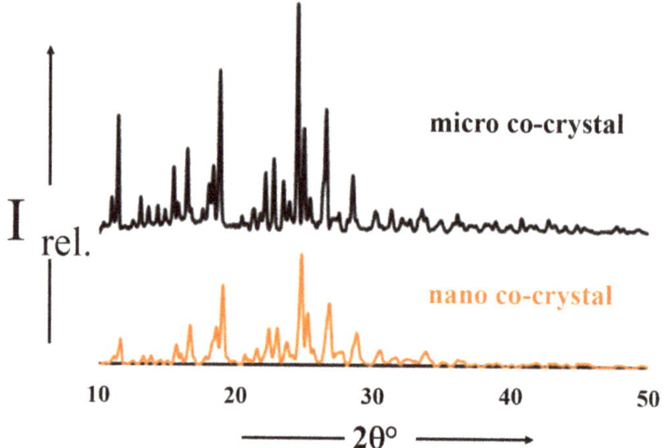

Figure 4. PXRD diffractograms for the micro (black) and nano co-crystals (orange).

3.2.5. Energy Dispersive X-ray Scanning Electron Microscopy (EDX-SEM)

The elemental composition of the nano and micro co-crystal indicates an identical percent composition of the elements present. The summary of elemental composition is summarized in Table 2. The data shows that the micro and nano co-crystal have the same elemental composition.

Table 2. Elemental composition of the micro and nano co-crystal.

Element	Micro Co-Crystal	Nano Co-Crystal
	Atomic %	Atomic %
C_K	48.26 ± 0.52	49.67 ± 1.21
N_K	20.75 ± 0.87	20.81 ± 0.94
O_K	29.96 ± 0.73	28.53 ± 1.02
S_K	1.03 ± 0.09	1.00 ± 0.03

3.3. Surfactant-Coated Co-Crystals

Nano co-crystals were initially synthesized without the use of surfactants and then with surfactants to identify the effect of surfactant addition on crystal production. Generally, the use of surfactants yielded co-crystals in the nanometer size range. The trend observed shows optimal stabilization was achieved when electrostatic stabilization is prevalent. Non-ionic surfactants with high hydrophilic-lipophilic balance (HLB) values exhibited low steric stabilization concentrations due to the presence of relatively large hydrophilic heads in the surfactant. PS, PDI and ZP data for surfactant free NCC have been added to Table 3 for comparative purposes. The data reported for PS is intensity distribution and z-average.

Table 3. Summary of results from general factorial experiments.

Std. Run	Run No.	Surfactant	Concentration % *w/v*	PS nm	PDI	ZP mV
8	1	TPGS 1000	1	200.6 ± 28.91	0.467 ± 0.077	−2.57 ± 0.63
2	2	SDS	0.5	1099 ± 166.10	0.811 ± 0.051	−18.2 ± 2.35
11	3	Span 80	2	351.5 ± 21.19	0.288 ± 0.078	−4.2 ± 1.22
10	4	SDS	2	182.1 ± 11.60	0.331 ± 0.086	−42.5 ± 3.41
7	5	Span 80	1	736.7 ± 77.15	0.663 ± 0.022	−7.1 ± 1.13
5	6	Tween 80	1	356 ± 42.09	0.357 ± 0.008	−1.04 ± 0.35
9	7	Tween 80	2	360 ± 88.67	0.558 ± 0.093	−2.8 ± 0.12
4	8	TPGS 1000	0.5	520 ± 55.32	0.479 ± 0.072	−3.08 ± 0.95
12	9	TPGS 1000	2	261.2 ± 19.94	0.483 ± 0.043	−1.57 ± 0.22
3	10	Span 80	0.5	1054 ± 224.67	1.000 ± 0.000	1.8 ± 0. 84
1	11	Tween 80	0.5	299 ± 40.40	0.36 ± 0.089	−6.2 ± 1.98
6	12	SDS	1	189.3 ± 2.65	0.323 ± 0.094	−28.2 ± 4.61
N/A	N/A	N/A	N/A	1593 ± 148.32	0.751 ± 0.063	−6.86 ± 1.04

3.3.1. Particle Size (PS) and Polydispersity Index (PDI)

The resultant particle size and PDI of the NCC was smaller when surfactant was used during the sonication phase of the production process. The interaction plots of PS and PDI using different surfactants and concentrations are depicted in Figures 5 and 6.

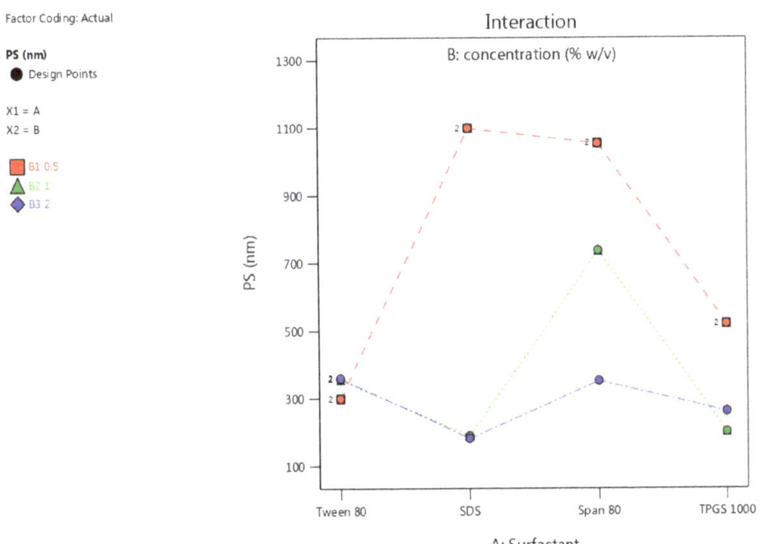

Figure 5. PS of nano co-crystals using different surfactants at concentrations of 0.5% *w/v* (red), 1% *w/v* (green) and 2% *w/v* (blue).

The reduction in particle size observed was greatest when SDS was used, due the electrostatic stabilization achieved with this stabilizer. At lower concentrations, higher molecular weight surfactants exhibited sufficient steric stabilization. For instance, Tween® 80 exhibited sufficient steric stabilization to produce crystals with a desirable PS and PDI, which can be attributed to the presence of a long hydrophobic chain and a considerably large hydrophilic head. It results in a reduction in PS and a low PDI indicating a narrow particle size distribution was achieved. The result is that fewer molecules of the surfactant are required to shield nucleating nano co-crystals from the balance of the materials in solution [25,59]. As the concentration of Tween® 80 increased, the PDI and PS of the nano co-crystals increased, and this was attributed to agglomeration behavior of the crystals. Of the non-ionic surfactants used, TPGS 1000 produced crystals with the smallest size at intermediate concentrations with a PDI < 0.500 at all concentrations tested. The larger polyethylene glycol molecules have more polar heads and result in stabilization at lower concentrations due to a larger shielding effect while also forming a monolayer around crystals. The nano co-crystals stabilized using Span® 80 exhibited a desirable PS and PDI as the concentration approached 2% *w/v*. Due to the presence of a smaller hydrophilic head, Span® 80 requires a much larger concentration to achieve the same CQA when compared to TPGS 1000 and Tween® 80. However, below the critical micelle concentration, ionic surfactants are better stabilizers than non-ionic compounds since ionic surfactants impart a surface charge to crystals, which results in electrical repulsion and better stabilization [60], as was evidenced for NCC stabilized using SDS.

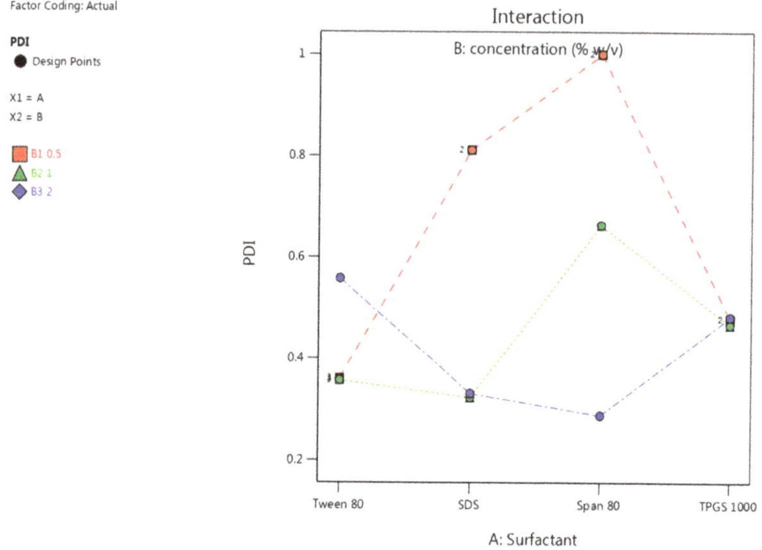

Figure 6. PDI of nano co-crystals using different surfactants at concentrations of 0.5% *w/v* (red), 1% *w/v* (green) and 2% *w/v* (blue).

The general trend depicted by the interaction plot for PDI in Figure 6 shows that an increase in concentration results in a lower PDI. As the concentration of each surfactant approached 2% *w/v* (blue line) the PDI values were lowest. This is attributable to the larg amount of available surfactant to prevent growth at the surface of several nuclei in solution resulting in a more uniform distribution of particle sizes. In addition to affecting the steric layer thickness, molecular weight and chain length; viscosity is also affected [60]. The higher the molecular weight, the higher the viscosity. In the final product, high viscosity can enhance the stability against aggregation [61]. This suggests that the molecular weight of the surfactants not only reduces particle or crystal size and PDI, but could potentially enhance stability.

3.3.2. Zeta Potential

The Zeta potential (ZP) was primarily affected by the presence of SDS, while the non-ionic surfactants resulted in a near neutral ZP. The rapid adsorption of negatively charged alkyl chains of SDS onto the surfaces of NCC, results in a negative charge at the surface [62]. The low ZP for the SDS stabilized nano-suspension follows the DLVO model [63]. The SO_4^{-2} anionic head group of SDS is adsorbed onto the surface of the NCC, resulting in a negative charge at the inner Helmholtz plane (IHP). As adsorption continues, the Langmuir adsorption isotherm potential of IHP is increased and eventually leads to a lower ZP in the original dispersion medium, when compared to that in water [63].

The non-ionic surfactants offer no electrostatic stabilization and therefore the resultant ZP is close to neutral. It is therefore, expected that the nano co-crystals stabilized with SDS would exhibit better stability in solution when compared to NCC stabilized with non-ionic surfactants [64,65]. The impact of surfactant type and concentration used, on the ZP is shown in Figure 7.

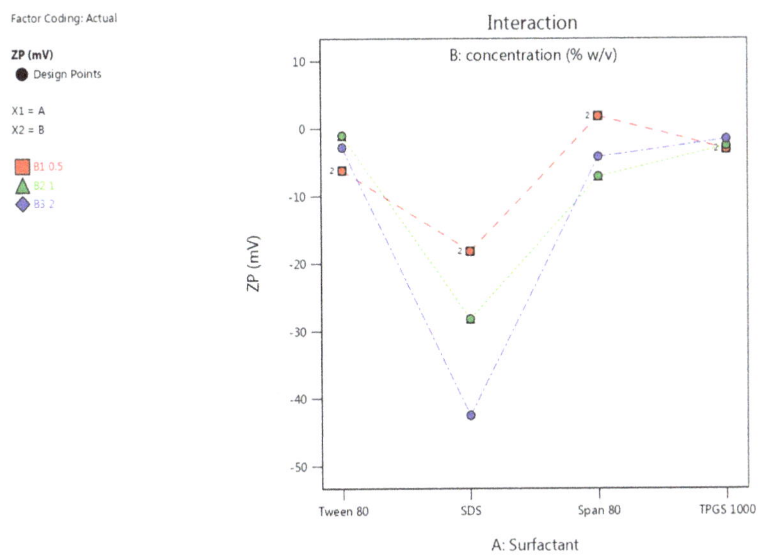

Figure 7. ZP of nano co-crystals using different surfactants at concentrations of 0.5% *w/v* (red), 1% *w/v* (green) and 2% *w/v* (blue).

4. Conclusions

3TC and AZT nano co-crystals were successfully prepared using a sonochemical synthesis approach. The use of different surfactants during the sonication phase of preparation resulted in the formation of crystals of reduced size, a reasonably narrow particle size distribution and in the case of the anionic surfactant, SDS, surface charge reduction. The use of surfactants to coat NCC may also achieve different purposes such as stealth and targeting capabilities, as these surfactant molecules are capable of interacting with specific substrates in biological systems [66–68].

All surfactants investigated exhibited concentration-dependent stabilization, resulted in the formation of nano co-crystals, and offered further stabilization characteristics. Stabilization was established to be concentration-dependent, based on the size of the hydrophilic head group of the surfactant used and surface charge induction. Surfactants with larger hydrophilic heads and thus larger HLB values, exhibited effective stabilization at lower concentration as observed by the ability of Tween® 80 and TPGS 1000 to produce co-crystals in the nanometer size range when used at low and medium concentration.

The PDI of the NCC produced also exhibited a direct relationship to the concentration of surfactant used to achieve stabilization. It is clear that surfactants act as a steric barrier to crystal growth generated by precipitation and as a result, nucleation rate rather than growth rate increases with increasing surfactant concentration, resulting in smaller dimensions and narrow size distributions [69,70].

The use of the anionic surfactant SDS produced crystals with a low ZP. The contribution to ZP reduction appears to be an antagonistic concentration-dependent relationship with the anionic surfactant SDS. An increase in SDS concentration resulted in lower ZP which would, consequently, result in an expected increase in solution stability of the technology.

These results suggest that the use of any of these surfactants would result in the production of co-crystals in the nanometer size range, provided that the correct concentration of surfactant is used while only the surfactant SDS would produce nano co-crystals that meet all the CQA criteria set prior to commencing these experiments. The findings of this research could prove useful in overcoming the low bioavailability of AZT and potentially provide a dosage form capable of delivering AZT and 3TC to HIV reservoirs, thereby potentially reducing the side effect profile associated with ARV treatment. In addition, the potential to produce a long-acting and -circulating ARV regimen is a possibility.

Investigations into the use of combinations of surfactants for the production of nano co-crystals to evaluate the possibility of producing specific/targeted CQA in co-crystals is ongoing in our laboratory.

Author Contributions: R.B.W. conceptualized and supervised the project. B.A.W. performed the experiments, analyzed the data and wrote the article. V.J.S. contributed to the conceptualization, supervision, bibliography research and proofreading of the manuscript. All authors have read and agreed to the published version of the manuscript.

Funding: This research was not funded with an external research grant.

Acknowledgments: The authors acknowledge the National Research Foundation (BAW) for a bursary, the Research Committee of Rhodes University (RBW) and Rhodes University Sandisa Imbewu fund (VJS) for financial assistance.

Conflicts of Interest: The authors declare no conflict of interest.

References

1. UNAIDS. *Global HIV Statistics*; UNAIDS: Geneva, Switzerland, 2019.
2. Subramanian, S. Manifestations of noncovalent bonding in the solid state. 6.' [~,(cyclam)]~+ (cyclam = 1,4,8,11 mtetraazacyclotetra- decane) as a template for crystal engineering of network hydrogen- bonded solids. *Can. J. Chem.* **1995**, *73*, 414–424. [CrossRef]
3. Blagden, N.; de Matas, M.; Gavan, P.T.; York, P. Crystal engineering of active pharmaceutical ingredients to improve solubility and dissolution rates. *Adv. Drug Deliv. Rev.* **2007**, *59*, 617–630. [CrossRef] [PubMed]
4. Aakeroy, C.B.; Aakeroy, A.; Sinha, A.S. *Co-Crystals: Introduction and Scope*; Royal Society of Chemistry: London, UK, 2018; Volume 11.
5. Bolton, O.; Matzger, A.J. Improved stability and smart-material functionality realized in an energetic cocrystal. *Angew. Chem. Int. Ed.* **2011**, *50*, 8960–8963. [CrossRef] [PubMed]
6. Brittain, H.G. Pharmaceutical cocrystals: The coming wave of new drug substances. *J. Pharm. Sci.* **2013**, *102*, 311–317. [CrossRef] [PubMed]
7. Brittain, H.G. Cocrystal Systems of Pharmaceutical Interest: 2010. *Cryst. Growth Des.* **2011**, *36*, 361–381. [CrossRef]
8. Sekhon, B. Pharmaceutical co-crystals—A review. *ARS Pharm.* **2009**, *150*, 99–117.
9. Gao, Y.; Zu, H.; Zhang, J. Enhanced dissolution and stability of adefovir dipivoxil by cocrystal formation. *J. Pharm. Pharmacol.* **2011**, *63*, 483–490. [CrossRef]
10. Yadav, A.; Shete, A.; Dabke, A.; Kulkarni, P.; Sakhare, S. Co-crystals: A novel approach to modify physicochemical properties of active pharmaceutical ingredients. *Indian J. Pharm. Sci.* **2009**, *71*, 359. [CrossRef]
11. Ahire, E.; Thakkar, S.; Darshanwad, M.; Misra, M. Parenteral nanosuspensions: A brief review from solubility enhancement to more novel and specific applications. *Acta Pharm. Sin. B* **2018**, *8*, 733–755. [CrossRef]
12. Trask, A.V. An overview of pharmaceutical cocrystals as intellectual property. *Mol. Pharm.* **2007**, *4*, 301–309. [CrossRef]
13. Merisko-Liversidge, E.; Liversidge, G.G. Nanosizing for oral and parenteral drug delivery: A perspective on formulating poorly-water soluble compounds using wet media milling technology. *Adv. Drug Deliv. Rev.* **2011**, *63*, 427–440. [CrossRef] [PubMed]
14. Merisko-Liversidge, E.; Liversidge, G.G.; Cooper, E.R. Nanosizing: A formulation approach for poorly-water-soluble compounds. *Eur. J. Pharm. Sci.* **2003**, *18*, 113–120. [CrossRef]
15. Sinha, B.; Muller, R.H.; Moschwitzer, J.P. Bottom-up approaches for preparing drug nanocrystals: Formulations and factors affecting particle size. *Int. J. Pharm.* **2013**, *8*, 384–392. [CrossRef] [PubMed]
16. De Waard, H.; Frijlink, H.W.; Hinrichs, W.L.J. Bottom-up preparation techniques for nanocrystals of lipophilic drugs. *Pharm. Res.* **2011**, *28*, 1220–1223. [CrossRef]

17. Xia, D.; Quan, P.; Piao, H.; Piao, H.; Sun, S.; Yin, Y.; Cui, F. Preparation of stable nitrendipine nanosuspensions using the precipitation-ultrasonication method for enhancement of dissolution and oral bioavailability. *Eur. J. Pharm. Sci.* **2010**, *40*, 325–334. [CrossRef]

18. Junyaprasert, V.B.; Morakul, B. Nanocrystals for enhancement of oral bioavailability of poorly water-soluble drugs. *Asian J. Pharm. Sci.* **2015**, *10*, 13–23. [CrossRef]

19. Lu, Y.; Wang, Z.H.; Li, T.; McNally, H.; Park, K.; Sturek, M. Development and evaluation of transferrin-stabilized paclitaxel nanocrystal formulation. *J. Control. Release* **2014**, *176*, 76–85. [CrossRef]

20. Sander, J.R.G.; Bučar, D.K.; Henry, R.F.; Zhang, G.G.Z.; MacGillivray, L.R. Pharmaceutical nano-cocrystals: Sonochemical synthesis by solvent selection and use of a surfactant. *Angew. Chem. Int. Ed.* **2010**, *49*, 7284–7288. [CrossRef]

21. Macgillivray, L.R.; Sander, J.R.; Bucar, D.K.; Elacqua, E.; Zhang, G.; Henry, R. Sonochemical synthesis of nano-cocrystals Boronic ester-based adducts. *Proc. Meet. Acoust.* **2013**, *19*, 45090.

22. Pi, J.; Wang, S.; Li, W.; Kebebe, D.; Zhang, Y.; Zhang, B.; Qi, D.; Guo, P.; Li, N.; Liu, Z. A nano-cocrystal strategy to improve the dissolution rate and oral bioavailability of baicalein. *Asian J. Pharm. Sci.* **2019**, *14*, 154–164. [CrossRef]

23. International Organization for Standardization. *ISO/TS 80004-2—Nanotechnologies—Vocabulary—Part 2:Nano-objects 2015*; International Organization for Standardization: Geneva, Switzerland, 2015.

24. Malamatari, M.; Taylor, K.M.G.; Malamataris, S.; Douroumis, D.; Kachrimanis, K. Pharmaceutical nanocrystals: Production by wet milling and applications. *Drug Discov. Today* **2018**, *23*, 534–547. [CrossRef] [PubMed]

25. Morsy, S.M.I. Role of surfactants in nanotechnology and their applications. *Int. J. Curr. Microbiol. Appl. Sci.* **2014**, *3*, 237–260.

26. Karashima, M.; Kimoto, K.; Yamamoto, K.; Kojima, T.; Ikeda, Y. A novel solubilization technique for poorly soluble drugs through the integration of nanocrystal and cocrystal technologies. *Eur. J. Pharm. Biopharm.* **2016**, *107*, 142–150. [CrossRef] [PubMed]

27. Teeranachaideekul, V.; Junyaprasert, V.B.; Souto, E.B.; Müller, R.H. Development of ascorbyl palmitate nanocrystals applying the nanosuspension technology. *Int. J. Pharm.* **2008**, *354*, 227–234. [CrossRef] [PubMed]

28. Hou, Y.; Shao, J.; Fu, Q.; Li, J.; Sun, J.; He, Z. Spray-dried nanocrystals for a highly hydrophobic drug: Increased drug loading, enhanced redispersity, and improved oral bioavailability. *Int. J. Pharm.* **2017**, *516*, 372–379. [CrossRef]

29. De Smet, L.; Saerens, L.; De Beer, T.; Carleer, R.; Adriaensens, P.; Van Bocxlaer, J.; Vervaet, C.; Remon, J.P. Formulation of itraconazole nanococrystals and evaluation of their bioavailability in dogs. *Eur. J. Pharm. Biopharm.* **2014**, *87*, 107–113. [CrossRef]

30. Ye, Y.; Zhang, X.; Zhang, T.; Wang, H.; Wu, B. Design and evaluation of injectable niclosamide nanocrystals prepared by wet media milling technique. *Drug Dev. Ind. Pharm.* **2015**, *41*, 1416–1424. [CrossRef]

31. Blasi, P.; Giovagnoli, S.; Schoubben, A.; Ricci, M.; Rossi, C. Solid lipid nanoparticles for targeted brain drug delivery. *Adv. Drug Deliv. Rev.* **2007**, *59*, 454–477. [CrossRef]

32. Fernandes, A.R.; Ferreira, N.R.; Fangueiro, J.F.; Santos, A.C.; Veiga, F.J.; Cabral, C.; Silva, A.M.; Souto, E.B. Ibuprofen nanocrystals developed by 22 factorial design experiment: A new approach for poorly water-soluble drugs. *Saudi Pharm. J.* **2017**, *25*, 1117–1124. [CrossRef]

33. Vuddanda, P.R.; Montenegro-Nicolini, M.; Morales, J.O.; Velaga, S. Effect of surfactants and drug load on physico-mechanical and dissolution properties of nanocrystalline tadalafil-loaded oral films. *Eur. J. Pharm. Sci.* **2017**, *109*, 372–380. [CrossRef]

34. Guo, Y.; Luo, J.; Tan, S.; Otieno, B.O.; Zhang, Z. The applications of Vitamin E TPGS in drug delivery. *Eur. J. Pharm. Sci.* **2013**, *49*, 175–186. [CrossRef] [PubMed]

35. Gupta, S.; Samanta, M.K.; Raichur, A.M. Dual-Drug Delivery System Based on In Situ Gel-Forming Nanosuspension of Forskolin to Enhance Antiglaucoma Efficacy. *AAPS PharmSciTech* **2010**, *11*, 322–335. [CrossRef]

36. Dash, P.K.; Gendelman, H.E.; Roy, U.; Balkundi, S.; Mosley, R.L.; Gelbard, H.A.; Mcmillan, J.; Gorantla, S.; Poluektova, L.Y. Long-acting NanoART Elicits Potent Antiretroviral and Neuroprotective Responses in HIV-1 Infected Humanized Mice. *AIDS* **2012**, *26*, 2135–2144. [CrossRef] [PubMed]

37. Mahesh, K.V.; Singh, S.K.; Gulati, M. A comparative study of top-down and bottom-up approaches for the preparation of nanosuspensions of glipizide. *Powder Technol.* **2014**, *256*, 436–449. [CrossRef]

38. Sohn, J.S.; Yoon, D.S.; Sohn, J.Y.; Park, J.S.; Choi, J.S. Development and evaluation of targeting ligands surface modified paclitaxel nanocrystals. *Mater. Sci. Eng. C* **2017**, *72*, 228–237. [CrossRef] [PubMed]

39. Rege, B.D.; Kao, J.P.Y.; Polli, J.E. Effects of nonionic surfactants on membrane transporters in Caco-2 cell monolayers. *Eur. J. Pharm. Sci.* **2002**, *16*, 237–246. [CrossRef]

40. Liu, M.; Hong, C.; Li, G.; Ma, P.; Xie, Y. The generation of myricetin-nicotinamide nanococrystals by top down and bottom up technologies. *Nanotechnology* **2016**, *27*, 395601. [CrossRef]

41. Bučar, D.-K.; Macgillivray, L.R. Preparation and Reactivity of Nanocrystalline Cocrystals Formed via Sonocrystallization. *J. Am. Chem. Soc.* **2007**, *129*, 32. [CrossRef]

42. Bhatt, P.M.; Azim, Y.; Thakur, T.S.; Desiraju, G.R. Co-Crystals of the Anti-HIV Drugs Lamivudine and Zidovudine. *Cryst. Growth Des.* **2009**, *9*, 951–957. [CrossRef]

43. Lupin Limited. *WO 2009/116055 Al 2009*; Lupin Limited: Mumbaid, India, 2009.

44. Peltonen, L.; Hirvonen, J. Pharmaceutical nanocrystals by nanomilling: Critical process parameters, particle fracturing and stabilization methods. *J. Pharm. Pharmacol.* **2010**, *62*, 1569–1579. [CrossRef]

45. Müllertz, A.; Perrie, Y.; Rades, T. *Advances in Delivery Science and Technology: Analytical Techniques in the Pharmaceutical Sciences*; Rathbone, M.J., Ed.; Springer Science and Business Media LLC: New York, NY, USA, 2016; ISBN 9781493940271.

46. Ebnesajjad, S. *Surface and Material Characterization Techniques*; Ebnesajjad, S., Ed.; Andrew William Applied Science Publishers: Oxford, UK, 2011; ISBN 978-1-4377-4461-3.

47. Brink, G. Infrared Studies of Water in Crystalline Hydrates: Ba(ClO3)2·H2O. *Appl. Spectrosc.* **1976**, *30*, 630–631. [CrossRef]

48. Falk, M.; Huang, C.-H.; Knop, O. Infrared Spectra of Water in Crystalline Hydrates: KSnCl 3H$_2$O, an Untypical Monohydrate. *Can. J. Chem.* **2006**, *52*, 2928–2931. [CrossRef]

49. Pereira, B.G.; Vianna-Soares, C.D.; Righi, A.; Pinheiro, M.V.B.; Flores, M.Z.S.; Bezerra, E.M.; Freire, V.N.; Lemos, V.; Caetano, E.W.S.; Cavada, B.S. Identification of lamivudine conformers by Raman scattering measurements and quantum chemical calculations. *J. Pharm. Biomed. Anal.* **2007**, *43*, 1885–1889. [CrossRef] [PubMed]

50. Du, Y.; Zhang, H.; Xue, J.; Tang, W.; Fang, H.; Zhang, Q.; Li, Y.; Hong, Z. Vibrational spectroscopic study of polymorphism and polymorphic transformation of the anti-viral drug lamivudine. *Spectrochim. Acta Part A Mol. Biomol. Spectrosc.* **2015**, *137*, 1158–1163. [CrossRef] [PubMed]

51. Mircescu, N.E.; Varvescu, A.; Herman, K.; Chiş, V.; Leopold, N. Surface-enhanced Raman and DFT study on zidovudine. *Spectroscopy* **2011**, *26*, 311–315. [CrossRef]

52. Palermo, E.F.; Chiu, J. Critical review of methods for the determination of purity by differential scanning calorimetry *. *Thermochim. Acta* **1976**, *14*, 1–12. [CrossRef]

53. Lai, S.L.; Guo, J.Y.; Petrova, V.; Ramanath, G.; Allen, L.H. Size-Dependent Melting Properties of Small Tin Particles: Nanocalorimetric Measurements. *Phys. Rev. Lett.* **1996**, *77*, 99–102. [CrossRef]

54. Schmidt, M.; Kusche, R.; von Issendorff, B.; Haberland, H. Irregular variations in the melting point of size-selected atomic clusters. *Nature* **1998**, *393*, 238–240. [CrossRef]

55. Chogale, M.M.; Ghodake, V.N.; Patravale, V.B. Performance parameters and characterizations of nanocrystals: A brief review. *Pharmaceutics* **2016**, *8*, 26. [CrossRef]

56. Kocbek, P.; Baumgartner, S.; Kristl, J. Preparation and evaluation of nanosuspensions for enhancing the dissolution of poorly soluble drugs. *Int. J. Pharm.* **2006**, *312*, 179–186. [CrossRef]

57. Kamb, W.B. Theory of Preferred Crystal Orientation Developed. *J. Geol.* **1958**, *67*, 153–170. [CrossRef]

58. Lang, A.R. X-ray diffraction procedures for polycrystal-line and amorphous materials. *Acta Metall.* **1956**, *4*, 102. [CrossRef]

59. Dalvi, S.V.; Dave, R.N. Controlling Particle Size of a Poorly Water-Soluble Drug Using Ultrasound and Stabilizers in Antisolvent Precipitation. *Ind. Eng. Chem. Res.* **2009**, *48*, 7581–7593. [CrossRef]

60. Rachmawati, H.; Al Shaal, L.; Muller, R.H.; Keck, C.M.; Shaal, L.A.; Müller, R.H.; Keck, C.M. Development of curcumin nanocrystal: Physical aspects. *J. Pharm. Sci.* **2013**, *102*, 204–214. [CrossRef]

61. Tuomela, A.; Hirvonen, J.; Peltonen, L. Stabilizing agents for drug nanocrystals: Effect on bioavailability. *Pharmaceutics* **2016**, *8*, 16. [CrossRef]

62. Song, S.-H.; Koelsch, P.; Weidner, T.; Wagner, M.S.; Castner, D.G. Sodium Dodecyl Sulfate Adsorption onto Positively Charged Surfaces: Monolayer Formation With Opposing Headgroup Orientations. *Langmuir* **2013**, *29*, 1–24. [CrossRef]

63. Keck, C.M. *Cyclosporine Nanosuspensions: Optimised Size Characterisation and Oral Formulations*; Freie Universitat Berlin: Berlin, Germany, 2006.
64. Mitri, K.; Shegokar, R.; Gohla, S.; Anselmi, C.; Müller, R.H. Lutein nanocrystals as antioxidant formulation for oral and dermal delivery. *Int. J. Pharm.* **2011**, *420*, 141–146. [CrossRef]
65. Ige, P.P.; Baria, R.K.; Gattani, S.G. Fabrication of fenofibrate nanocrystals by probe sonication method for enhancement of dissolution rate and oral bioavailability. *Colloids Surf. B Biointerfaces* **2013**, *108*, 366–373. [CrossRef]
66. Muller, R.H.; Keck, C.M. Challenges and solutions for the delivery of biotech drugs—A review of drug nanocrystal technology and lipid nanoparticles. *J. Biotechnol.* **2004**, *113*, 151–170. [CrossRef]
67. Huang, X.; Peng, X.; Wang, Y.; Wang, Y.; Shin, D.M.; El-Sayed, M.A.; Nie, S. A reexamination of active and passive tumor targeting by using rod-shaped gold nanocrystals and covalently conjugated peptide ligands. *ACS Nano* **2010**, *4*, 5887–5896. [CrossRef]
68. Wang, T.; Qi, J.; Ding, N.; Dong, X.; Zhao, W.; Lu, Y.; Wang, C.; Wu, W. Tracking translocation of self-discriminating curcumin hybrid nanocrystals following intravenous delivery. *Int. J. Pharm.* **2018**, *546*, 10–19. [CrossRef] [PubMed]
69. Matteucci, M.E.; Hotze, M.A.; Johnston, K.P.; Williams, R.O. Drug nanoparticles by antisolvent precipitation: Mixing energy versus surfactant stabilization. *Langmuir* **2006**, *22*, 8951–8959. [CrossRef] [PubMed]
70. Dong, Y.; Ng, W.K.; Shen, S.; Kim, S.; Tan, R.B.H. Preparation and characterization of spironolactone nanoparticles by antisolvent precipitation. *Int. J. Pharm.* **2009**, *375*, 84–88. [CrossRef] [PubMed]

Article

Quality by Design Optimization of Cold Sonochemical Synthesis of Zidovudine-Lamivudine Nanosuspensions

Bwalya A. Witika [1], Vincent J. Smith [2] and Roderick B. Walker [1],*

[1] Division of Pharmaceutics, Faculty of Pharmacy, Rhodes University, Makhanda 6140, South Africa; bwawitss@gmail.com

[2] Department of Chemistry, Faculty of Science, Rhodes University, Makhanda 6140, South Africa; v.smith@ru.ac.za

* Correspondence: r.b.walker@ru.ac.za

Received: 3 March 2020; Accepted: 30 March 2020; Published: 17 April 2020

Abstract: Lamivudine (3TC) and zidovudine (AZT) are antiviral agents used to manage HIV/AIDS infection. The compounds require frequent dosing, exhibit unpredictable bioavailability and a side effect profile that includes hepato- and haema-toxicity. A novel pseudo one-solvent bottom-up approach and Design of Experiments using sodium dodecyl sulphate (SDS) and α-tocopheryl polyethylene glycol succinate 1000 (TPGS 1000) to electrosterically stablize the nano co-crystals was used to develop, produce and optimize 3TC and AZT nano co-crystals. Equimolar solutions of 3TC in surfactant dissolved in de-ionised water and AZT in methanol were rapidly injected into a vessel and sonicated at 4 °C. The resultant suspensions were characterized using a Zetasizer and the particle size, polydispersity index and Zeta potential determined. Optimization of the nanosuspensions was conducted using a Central Composite Design to produce nano co-crystals with specific identified and desirable Critical Quality Attributes including particle size (PS) < 1000 nm, polydispersity index (PDI) < 0.500 and Zeta potential (ZP) < −30mV. Further characterization was undertaken using Fourier Transform infrared spectroscopy, energy dispersive X-ray spectroscopy, differential scanning calorimetry, powder X-ray diffraction and transmission electron microscopy. In vitro cytotoxicity studies revealed that the optimized nano co-crystals reduced the toxicity of AZT and 3TC to HeLa cells.

Keywords: nano co-crystals; design of experiments; Quality by Design; crystal engineering; sonochemistry; HIV/AIDS; lamivudine; zidovudine

---

1. Introduction

Since the beginning of the HIV epidemic in the 1980s, 74.9 million people have been infected with the virus. In the same time period, an approximate total of 32 million people have died from AIDS-related illnesses [1].

According to a report by UNAIDS, approximately 37.9 million people globally were living with HIV at the end of 2018. Of these, 23.3 million were accessing antiretroviral therapy (ART). In addition, 1.7 million new HIV infections were reported in 2018. HIV/AIDS also accounted for 770,000 deaths from AIDS-related illnesses [1].

Co-crystals incorporate pharmaceutically acceptable guest molecules into a crystal together with an active pharmaceutical ingredient (API). Co-crystals have gained attention as alternate solid forms for drug development [2]. For pharmaceutical applications, co-crystallization is highly promising for tailoring the properties of Active Pharmaceutical Ingredients (API) and co-former couples to improve dissolution kinetics, the rate and extent of bioavailability, and the stability of compounds [3–6].

Multi-API co-crystals, which remain relatively unexplored solid forms of API, have potential relevance in the context of combination therapy for drug and product development [7]. AZT and 3TC are antiviral agents used in combination for the suppression and prevention of HIV/AIDS [3,8–11]. The 3TC·AZT·H$_2$O co-crystal has been synthesized by neat grinding, liquid-assisted grinding and solvent evaporation [12]. The co-crystal has the potential for co-delivery of both molecules.

Co-crystals can be produced in the nanometre range using top-down or bottom-up manufacturing approaches. The production of nanocrystals and nano co-crystals using bottom-up techniques generally involve the use of a solvent system(s) added to an anti-solvent system in the presence of ionic and/or non-ionic stabilisers. The mixing of drug solution(s) and anti-solvent is generally achieved with conventional mixing equipment such as magnetic and/or overhead stirrers fitted with an agitator blade [13]. In order to promote nucleation, sonic waves are introduced using a process now called, sonoprecipitation [13,14]. High-energy mechanical forces are involved when using top-down approaches, which are achieved using media milling (MM) (NanoCrystals®) or High Pressure Homogenization (IDD-P®, DissoCubes® and Nanopure®) to comminute large crystals [15,16].

The use of surfactants and polymers such as Tween ®, Span ®, hydroxypropyl methylcellulose (HPMC), pyrrolidone K30 and polyvinyl pyrrolidone as stabilizers has been explored in the synthesis of nanocrystals and nano co-crystals (NCC) [17–25].

Macrophages are carriers of HIV in humans [26–28] and targeting these cells by use of nanotechnological approaches has been proposed as an ideal option for more efficient treatment of HIV and halting the progression to AIDS [29–32]. Nanocrystals and nano co-crystals offer a potential route for targeting macrophages. Following their administration, nanocrystals and nano co-crystals are engulfed by phagocytic cells such as macrophages after recognition as foreign bodies. In phagocytic cells, nano-dimensional crystals dissolve slowly in phagolysosomes. Consequently, any payload might pass through phagolysosomal membranes and reach the cytoplasm of cells, following which diffusion from the cell down a concentration gradient occurs [15].

Cold-sonochemical co-crystallization as a process has been used successfully for the manufacture of nano co-crystals [14,33,34]. One-solvent systems involve dissolving the components in a single solvent and injecting the solution into an anti-solvent while applying ultrasonic energy [33], whereas in two-solvent systems, the components are dissolved in separate vehicles and injected into an anti-solvent. Preliminary studies suggested that neither of these approaches were suitable due to differences in the solubility of 3TC and AZT. Consequently, we developed a pseudo one-solvent system approach in which the components were dissolved in separate solvents that acted as anti-solvents for each other in situ [25].

The pseudo one-solvent cold-sonochemical approach [25] was used to synthesize and optimize co-crystals in the nanometer range with a specific Zeta potential that could potentially exploit the advantages of nanometer drug delivery systems for targeted drug delivery [35,36] as well as those of co-crystalline drug delivery systems such as enhanced solubility [37,38] and for combination therapy [39]. We have previously demonstrated the preparation of self-assembled electrosterically (both sterically and electrostatically) stabilized nano co-crystals using SDS as an electrostatic stabilizer and TPGS 1000 as a steric stabilizer using an appropriate anti-solvent [25].

A Design of Experiments (DoE) approach, specifically Central Composite Design (CCD) with the aid of Response Surface Methodology (RSM) was used to optimize formulation variables for the synthesis of 3TC and AZT nano co-crystals using cold-sonoprecipitation. Preliminary studies revealed that two independent factors viz., SDS and TPGS 1000 concentration were important.

The main purpose of these studies was to develop and optimize, using the principles of Quality by Design (QbD), a surface-modified nano co-crystal formulation, using pseudo one-solvent bottom-up cold-sonoprecipitation. The NCC will have the potential to deliver 3TC and AZT to target reservoirs of HIV in different tissues, with the potential of improving the side effect profile of each API by reducing the dose. The surfactants, TPGS 1000 and SDS were identified and reported previously [25] and the optimization of the formulation and cytotoxicity of the nano co-crystal is reported herein.

2. Materials and Methods

2.1. Materials

AZT and 3TC were purchased from China Skyrun Co. Ltd. (Taizhou, China). SDS and TPGS 1000 were purchased from Merck (Johannesburg, South Africa). HPLC-grade water was prepared by reverse osmosis using a RephiLe® Direct-Pure UP and RO water system Microsep® (Johannesburg, South Africa) fitted with a RephiDuo® H PAK de-ionization cartridge and a RephiDuo® PAK polishing cartridge. The water was filtered through a 0.22 µm PES high-flux capsule filter Microsep® (Johannesburg, South Africa) prior to use. HPLC-grade methanol (MeOH) was purchased from Honeywell Burdick and Jackson™ (Anatech Instruments, Johannesburg, South Africa).

2.2. Methods

2.2.1. Nano Co-Crystal Synthesis

NCC were synthesized using a cold-sonoprecipitation pseudo one-solvent method as previously described [25]. The batch size was approximately 1 g, specifically 534 mg of AZT and 458 mg of 3TC equivalent to 2 mmol of each API. 3TC was dissolved in 7 mL water and AZT in 6 mL MeOH. The surfactants were added to 3TC/water solution in concentrations as required by the CCD. The solutions were rapidly injected into a pre-cooled conical flask, and then incubated at 4 °C ± 2 °C in an ice bath. A sonication output of 50 kHz ± 6 kHz was applied to the solution for 20 min using a Branson® 8510E-MT ultrasonic bath (Branson Ultrasonics Corp., Danbury, CT, USA).

2.2.2. Nano Co-Crystal Optimization

A CCD design generated using version 11 Design Expert® software (Stat-Ease Inc., Minneapolis, MN, USA) was selected for the optimization of process variables for the synthesis of NCC using reduced-temperature sonoprecipitation. Preliminary studies revealed that two independent factors viz., SDS and TPGS 1000 concentration were critical [25]. The PS, PDI and ZP were monitored. The experiments conducted for the CCD are listed in Table 1. All experiments were performed in triplicate for each run using the levels suggested by the CCD, which also includes repeated runs at specific levels. Furthermore, the runs are conducted in a randomised manner to minimize any potential bias.

Table 1. Summary of CCD experiments for nano co-crystals (NCC) optimization.

Std. Run	Form. Code	SDS % w/v	TPGS 1000 % w/v
4	1	1.00	2.00
11	2	0.50	1.50
5	3	0.00	1.50
2	4	1.00	1.00
7	5	0.50	0.79
8	6	0.50	2.21
13	7	0.50	1.50
6	8	1.21	1.50
1	9	0.00	1.00
3	10	0.00	2.00
9	11	0.50	1.50
10	12	0.50	1.50
12	13	0.50	1.50

Sodium dodecyl sulphate (SDS), α-tocopheryl polyethylene glycol succinate (TPGS 1000).

2.2.3. Particle Size Analysis

The mean PS and PDI of NCC was determined using a Nano-ZS 90 Zetasizer (Malvern Instruments, Worcestershire, UK) with the instrument set in Photon Correlation Spectroscopy (PCS) mode. Approximately 30 µL of an aqueous dispersion of NCC was diluted with 10 mL HPLC-grade water prior to analysis. The sample was placed into a $10 \times 10 \times 45$ mm polystyrene cell and all measurements were performed in replicate ($n = 6$) at 25 °C using a standard 4 mW laser set at 633 nm at a scattering angle of 90°. Analysis of PCS data was undertaken using Mie theory with real and imaginary refractive indices set at 1.456 and 0.01.

2.2.4. Zeta Potential

The ZP of NCC was measured using a Nano-ZS 90 Zetasizer (Malvern Instruments, Worcestershire, UK) set in Laser Doppler Anemometry (LDA) mode ($n = 6$) at a wavelength of 633 nm. The samples were prepared for analysis as described in Section 2.2.2 and placed into folded polystyrene capillary cells prior to analysis.

2.2.5. Formulation Optimization

Numerical optimization was undertaken using version 11 Design Expert® software (Stat-Ease Inc.) with the aim of identifying an optimum formulation composition and associated formulation parameters that would ensure the production of nanosuspension formulations of optimum stability with the potential to target reservoirs of HIV in the body. Although different methods are used to achieve optimization of process and formulation parameters, the use of a numerical optimization is a comprehensive and effective approach for any continuous optimization process [40]. Numerical optimization locates a point in space that maximises the desirability function while modifying the characteristics of a target by adjusting the importance of that target [41]. The summary of formulation parameters used for the optimization of nano co-crystals are listed in Table 2.

Table 2. Summary of optimized formulation conditions for the manufacture of optimized nano co-crystal (OPT-NCC).

Formulation Variable	Optimized Condition
SDS concentration (X_1)	0.90% w/v
TPGS 1000 concentration (X_2)	1.40% w/v

2.3. Characterization of Optimized Nano Co-Crystal (OPT-NCC)

2.3.1. Differential Scanning Calorimetry

Approximately 4 mg of filtered and air-dried OPT-NCC (to constant mass) was placed into an aluminum pan and sealed. The pan was then placed directly into the furnace of a DSC 6000 PerkinElmer Differential Scanning Calorimeter (Waltham, MA, USA) and the data generated was analyzed using version 11 Pyris™ Manager Software (PerkinElmer). The temperature of the DSC was monitored with a computer and a controlled heating rate of 10 K/min was used for the analysis over the temperature range 30–150 °C. All DSC analyses were conducted in triplicate ($n = 3$) under a nitrogen atmosphere purged at a flow rate of 20 mL/min and the thermogram for the micro co-crystal (uncoated) was used as a reference.

2.3.2. Energy-Dispersive X-ray Spectroscopy Scanning Electron Microscopy

Elemental analysis was performed using a Vega® Scanning Electron Microscope (TESCAN, Brno, Czechia) fitted with an INCA PENTA FET. Approximately 1 mg of the OPT-NCC was dusted onto a graphite plate and the sample irradiated at an accelerated voltage of 20 kV ($n = 3$).

2.3.3. FTIR Spectroscopy

The IR absorption spectrum of the OPT-NCC was generated using a Model 100 Spectrum FTIR ATR Spectrophotometer (PerkinElmer) and analyzed using version 4.00 Peak® Spectroscopy software (Operant LLC, Burke, VA, USA). Approximately 5 mg powder was placed onto a diamond crystal and analyzed over the wavenumber range 4000–650 cm^{-1} at a rate of 4 cm^{-1} in replicate ($n = 5$) and the spectrum for the uncoated nano co-crystals was used for reference purposes.

2.3.4. Powder X-ray Diffraction (PXRD)

X-ray powder diffraction patterns were collected using a Bruker D8 Discover diffractometer (Bruker, Billerica, MA, USA) equipped with a proportional counter, Cu-Kα radiation of $\lambda = 1.5405$ Å and a nickel filter. Samples were placed onto a silicon wafer slide. Generator settings were 30 kV with a current of 40 mA used for the measurement. Data were collected ($n = 3$) in the range $2\theta = 10°$ to $50°$ at a scanning rate of 1.5° min^{-1} with a filter time-constant of 0.38 s per step and a slit width of 6.0 mm. The X-ray diffraction data were treated using evaluation curve fitting (Diffrac.Eva, version: V2.9.0.22, Bruker) software. Baseline correction was performed on each diffraction pattern by subtracting a spline function fitted to the curved background and the diffraction pattern of uncoated nano co-crystals was used for reference purposes.

2.3.5. Transmission Electron Microscopy (TEM)

TEM was used to visualize the shape of the NCC in the original aqueous dispersion. A drop of the aqueous dispersion was placed onto a copper grid with a carbon film and excess liquid was removed using Whatman® 110 filter paper (Whatman PLC, Maidstone, England), after which the sample was dried at room temperature (22 °C) for 24 h. The sample was visualised using a Zeiss Libra® 120 TEM (Carl Zeiss AG, Munich, Germany).

2.3.6. In Vitro Cytotoxicity Studies

HeLa (human cervix adenocarcinoma cells) (Cellonex) were cultured in Dulbecco's Modified Eagle Medium (DMEM)-(Lonza) supplemented with 10% w/w fetal calf serum and antibiotics (penicillin/streptomycin/amphotericin B) at 37 °C in a 5% CO_2 incubator. HeLa cells were transferred to 96-well plates at a cell density of 1×10^4 cells per well in 150 μL culture medium and grown overnight. A single concentration of 50 μM of the test compound was incubated with the cells for an additional 48 h, and cell viability in the wells assessed by adding 20 μL 0.54 mM resazurin in PBS for an additional 2–4 h. The numbers of cells surviving drug treatment were determined by reading resorufin fluorescence (excitation 560 nm, emission 590 nm) with a SpectraMax® M3 plate reader (Molecular Devices, San Jose, CA, USA). Fluorescence readings for the individual wells were converted to percent (%) cell viability relative to the average readings from untreated control wells. Plots of % cell viability vs. log(compound) were used to determine IC$_{50}$ values by non-linear regression using version 5.02 GraphPad Prism (GraphPad Holdings LLC, La Jolla, CA, USA).

The compounds investigated were the optimized NCC, the individual API and a physical mixture of the API in stoichiometric ratios identical to those used to produce the NCC.

3. Results

3.1. Optimization of Electrosteric NCC

A summary of the input variables used to optimize the manufacture of NCC identified using the CCD are summarized in Table 3.

Table 3. Summary of variables used and responses for NCC produced using CCD.

Summary of CCD Experiments for Optimisation				Responses for NCC Produced Using CCD		
Std. Run	Form. Code	SDS % w/v	TPGS % w/v	PS nm	PDI	ZP mV
4	1	1.00	2.00	451.9 ± 43.1	0.689 ± 0.003	−45.0 ± 3.9
11	2	0.50	1.50	446.8 ± 52.8	0.478 ± 0.034	−27.7 ± 1.2
5	3	0.00	1.50	705.6 ± 92.3	0.474 ± 0.092	−22.3 ± 0.9
2	4	1.00	1.00	475.9 ± 22.3	0.506 ± 0.033	−30.5 ± 2.6
7	5	0.50	0.79	630.1 ± 73.2	0.414 ± 0.045	−27.1 ± 1.4
8	6	0.50	2.21	304.1 ± 38.6	0.498 ± 0.087	−21.9 ± 2.3
13	7	0.50	1.50	436.4 ± 67.1	0.471 ± 0.011	−25.9 ± 3.1
6	8	1.21	1.50	404.9 ± 73.6	0.462 ± 0.038	−37.0 ± 0.3
1	9	0.00	1.00	859.4 ± 101.2	0.448 ± 0.088	−13.3 ± 3.7
3	10	0.00	2.00	656.9 ± 77.2	0.284 ± 0.018	−13.5 ± 2.2
9	11	0.50	1.50	424.9 ± 21.7	0.432 ± 0.028	−26.2 ± 1.6
10	12	0.50	1.50	423.7 ± 13.2	0.449 ± 0.049	−25.3 ± 0.7
12	13	0.50	1.50	430.7 ± 52.1	0.466 ± 0.009	−24.8 ± 2.7

Shaded cells reflect results that have met the critical quality attributes (CQA) [25]. PS, particle size; PDI, polydispersity index; ZP, Zeta potential.

3.1.1. Response Surface Quadratic Model for PS (Y_1)

The ANOVA results for the response surface quadratic model for PS are listed in Table 4. The model F value was 26.82 indicating that the model was significant. The model F-value is used to ascertain the utility of a model that the data has been fitted to and determine whether the data is best fitted by the model. The F-value is explained and unexplained variability and the larger the F-value, the more useful the model [42]. The particle size was influenced by SDS and TPGS 1000 concentration and an increase in % w/v content of each resulted in a reduction in particle size.

Table 4. ANOVA data for response surface quadratic model for PS.

Source	Sum of Squares	df	Mean Square	F-Value	*p*-Value Prob. > F
Model	2.71×10^5	5	54,111.2	26.82	0.0002
A-SDS Conc.	1.68×10^5	1	1.679×10^5	83.23	<0.0001
B-TGPS Conc.	59,087.81	1	59,087.8	29.29	0.0010
AB	7965.56	1	7965.56	3.95	0.0873
A^2	73,941.57	1	73,941.6	36.65	0.0005
B^2	6594.86	1	6594.86	3.27	0.1135
Residual	14,121.22	7	2017.32		
Lack of Fit	13,763.08	3	4587.69	51.24	0.0012
Pure Error	358.14	4	89.54		
Corr. Total	2.85×10^5	12			

Significant factors are reported in red.

The mean PS of the NCC fell between 304.1 and 859.4 nm. The impact of SDS on the size of NCC resulted in a large F-value of 83.23, indicating that this parameter had a significant impact on the size of NCC produced when compared to the other variable investigated. However, the TPGS 1000 content and the quadratic effect of the SDS concentration produced F-values of 29.29 and 36.65, respectively, indicating that these parameters had an intermediate, yet significant, impact on the resultant size of the NCC produced.

The three-dimensional (3D) response surface plot in which the impact of SDS and TPGS 1000 concentration on particle size is depicted in Figure 1. These data reveal that a synergistic relationship exists between the SDS and TPGS 1000 concentration on particle size. The surface plots reveal that relatively small NCC are produced when the amount of SDS used is high and TPGS 1000 content is constant. The same effect is observed when the SDS concentration is constant and TGPS 1000 concentration is increased. However, this effect is minimal when SDS concentrations are at a maximum of 1% w/v. These results are consistent with previous findings that reported a decrease in mean PS with increasing content of an electrostatic or steric surfactant [20,43–45]. The desired mean particle size can be produced by a manipulation of the combined effects of SDS and TPGS 1000 when producing NCC using this method.

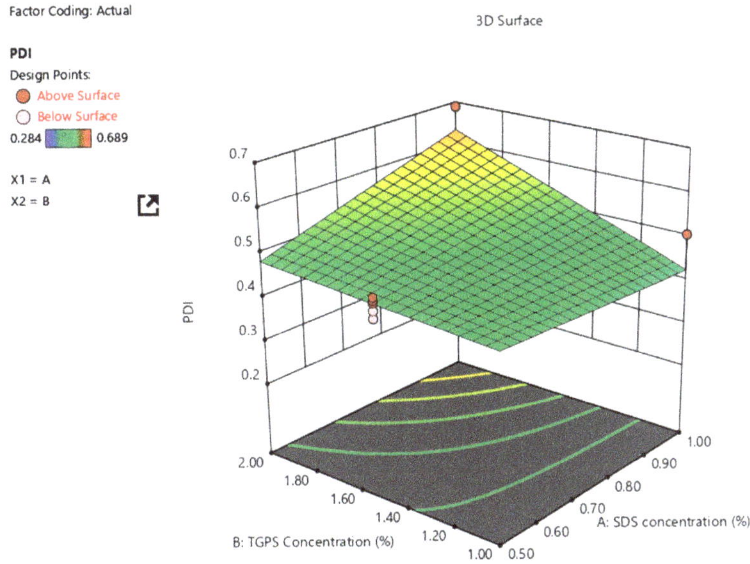

Figure 1. 3D response surface plot depicting the impact of SDS and TPGS 1000 concentration on particle size of the NCC.

3.1.2. Response Surface Model for PDI (Y_2)

The ANOVA results for the response surface linear model for PDI are listed in Table 5. The model F value was 6.37, indicating that the model was significant. Increasing the amount of SDS used increases the PDI.

Table 5. ANOVA data for response surface linear model for PDI.

Source	Sum of Squares	df	Mean Square	F-Value	*p*-Value Prob. > F
Model	0.061	3	0.02	6.37	0.0132
A-SDS Concentration	0.029	1	0.029	8.99	0.015
B-TGPS Concentration	2.37×10^{-3}	1	2.37×10^{-3}	0.74	0.412
AB	0.03	1	0.03	9.39	0.0135
Residual	0.029	9	3.21×10^{-3}		
Lack of Fit	0.027	5	5.496×10^{-3}	15.9	0.0096
Pure Error	1.38×10^{-3}	4	3.457×10^{-3}		
Corr. Total	0.09	12			

Significant factors are reported in red.

The 3D response surface plot in Figure 2 depicts the impact of SDS concentration on PDI and that a linear correlation exists. A reduction of TPGS 1000 concentration appears to increase the PDI for the NCC when the SDS concentration is kept constant, but only marginally.

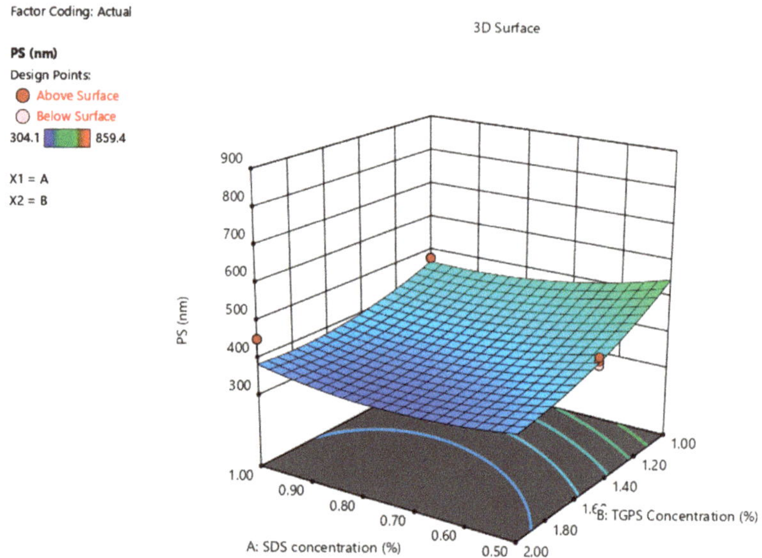

Figure 2. 3D response surface plot depicting the impact of SDS and TPGS 1000 concentration on the PDI of the NCC.

3.1.3. Response Surface Model for ZP (Y_3)

The ANOVA results for the response surface quadratic model for ZP are listed in Table 6. The model F value was 17.56, indicating that the model was significant. Increasing the amount of SDS used results in a decrease in ZP.

Table 6. ANOVA data for response surface linear model for ZP.

Source	Sum of Squares	df	Mean Square	F-Value	*p*-Value Prob. > F
Model	666.17	2	333.08	17.56	0.0005
A-SDS Concentration	659.42	1	659.42	34.76	0.0002
B-TGPS Concentration	6.75	1	6.75	0.36	0.5642
Residual	189.72	10	18.97		
Lack of Fit	184.85	6	30.81	25.32	0.0038
Pure Error	4.87	4	1.22		
Corr. Total	855.89	12			

Significant factors are reported in red.

The 3D response surface plot in Figure 3 depicts the impact of SDS concentration on the ZP and that a linear correlation exists between SDS content and ZP. The concentration of TPGS 1000 appears to have an insignificant impact on the ZP of the NCC. The effect of SDS on ZP is consistent with that previously reported for SDS-stabilized nanosuspensions [46–50].

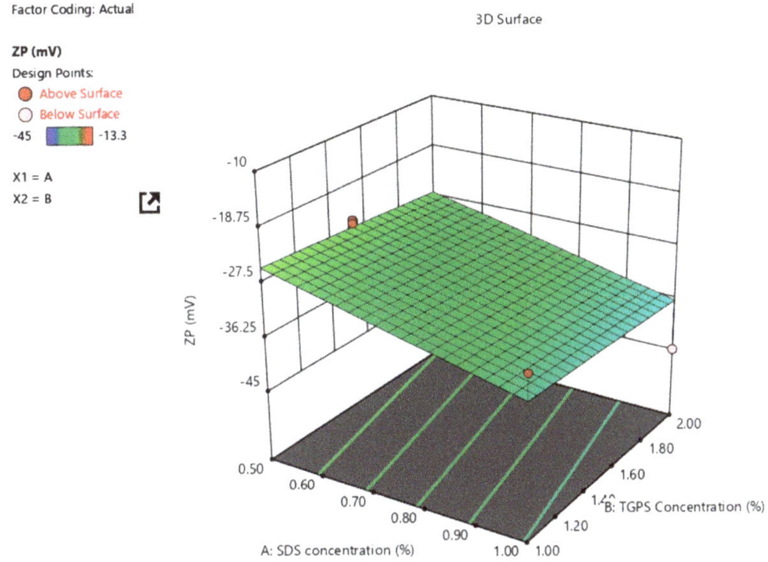

Figure 3. 3D response surface plot depicting the impact of SDS and TPGS 1000 concentration on ZP of the NCC.

3.1.4. Formulation Optimization

The desired target level for each input variable and associated responses PS, PDI and ZP are summarised in Table 7. The desirability of the model generated was 1.000, indicating that the optimum conditions were located in the desirability zone. The optimum formulation composition for the manufacture OPT-NCC and is summarised in Table 7.

Table 7. Formulation variables and associated responses.

Formulation Variables		Formulation Responses			Desirability
SDS % w/v	TPGS % w/v	PS Nm	PDI	ZP mV	
0.90	1.40	393.08	0.499	−33.47	1.000

The responses generated using the NCC formulation developed and manufactured according to the optimum composition are summarised in Table 8, in addition to the experimental and predicted responses with the corresponding percent prediction error.

Table 8. Summary of results for the NCC produced using optimum formulation parameters.

Response	Predicted Value	Experimental Value	% Predicted Error
PS nm	393.08	332.9 ± 42.85	18.08
PDI	0.499	0.474 ± 0.040	5.27
ZP mV	−33.47	−34.6 ± 5.56	3.27

3.2. Characterization of OPT-NCC

3.2.1. Differential Scanning Calorimetry

DSC was used to investigate whether polymorphic changes had occurred when manufacturing the NCC and the resultant thermogram is depicted in Figure 4. The thermogram revealed a melting endotherm for the OPT-NCC with a T_{peak} at 94.3 °C and for the uncoated NCC at T_{peak} at 103.1 °C. The reduction in melting point is likely due to particle size reduction [51–53]. The NCC has a narrow melting endotherm, indicating that crystallinity is retained during the production process. Tabulated melting point data are reported in Table S1 and the thermograms for the components and the NCC are depicted in Figure S1 in the Supplementary Information.

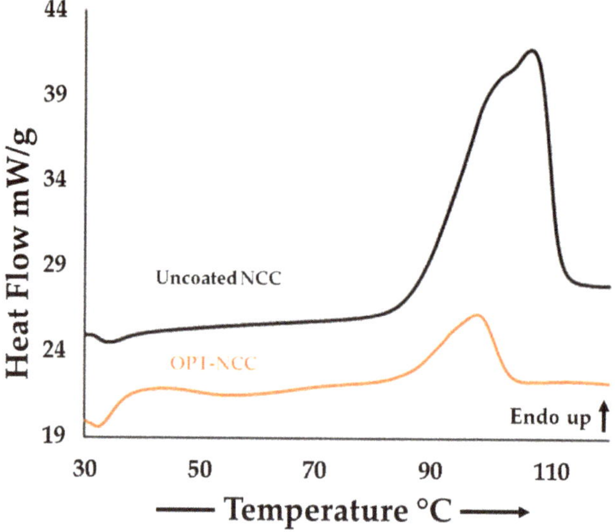

Figure 4. DSC thermograms depicting the melting endotherm for the uncoated (black) and OPT-NCC (orange).

3.2.2. SEM-EDX

Elemental analysis indicates the presence of elemental sodium in SDS/TPGS stabilized NCC. The OPT-NCC have lower oxygen content when compared to uncoated NCC and the overall weight ratio is lower due to the presence of elemental sodium and high sulphur content from the combination of stabilizers used. A summary of the EDX measurements is listed in Table 9 and depicted in Figure 5.

Table 9. Summary of elemental analysis for the physical mixture, NCC-OPT and co-crystal.

Element	OPT-NCC	Uncoated NCC
	Atomic %	Atomic %
C_k	56.97 ± 2.18	49.67 ± 1.21
N_k	13.99 ± 0.94	20.81 ± 0.94
O_k	25.87 ± 1.22	28.53 ± 1.02
Na_k	0.26 ± 0.13	-
S_k	2.97 ± 0.34	1.00 ± 0.03

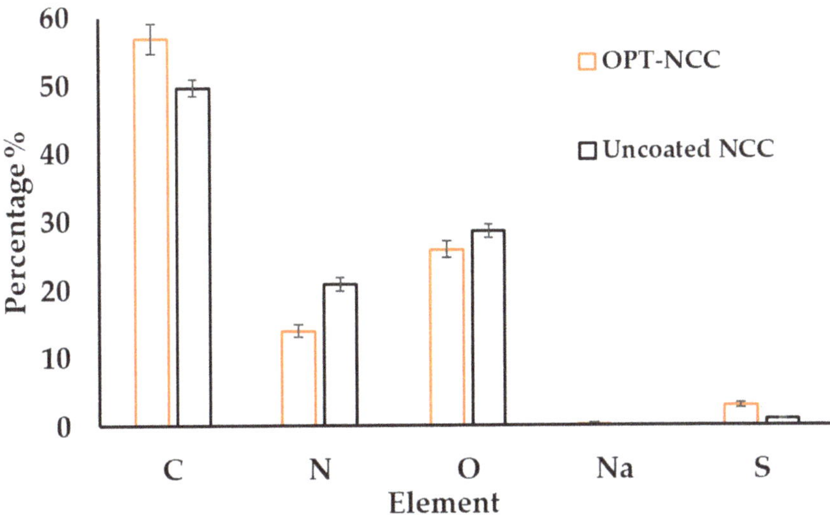

Figure 5. Elemental analysis of uncoated (black) and OPT-NCC (orange).

3.2.3. FTIR Spectroscopy

The FTIR spectra depicted in Figure 6 show peaks at 3530 cm^{-1} (*), which are characteristic of the hydrogen-bonded water in the crystal for both coated and uncoated nano co-crystals, [54,55]. The stretching band at 1635 cm^{-1} (@) is due to the carbonyl moiety (O=C–NR$_2$) and is characteristic for 3TC and AZT. It partially overlaps with the N–H (bending) band at 1607 cm^{-1} (→). The stretching vibration of the imine group (R$_2$-C=NR) is observed at 1648 cm^{-1} (#). Characteristic bands for AZT are observed at 2170 cm^{-1} (‡) and 1652 cm^{-1}, due to –N$_3$ and –N–H stretching vibrations. A comparison of the wavenumbers for the co-crystal and the OPT-NCC are reported and plotted in Table S2 and Figure S2 respectively and the spectra for the NCC, AZT and 3TC are depicted in Figure S3 in the Supplementary Information.

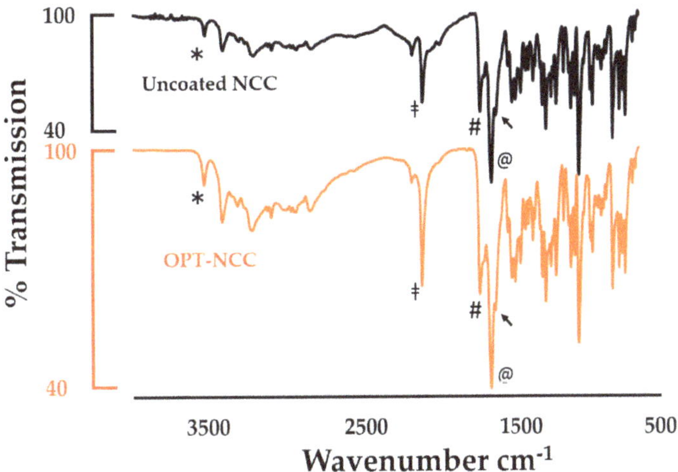

Figure 6. FTIR absorption spectra of the uncoated NCC (black) and the OPT-NCC (orange).

3.2.4. PXRD

The diffractograms depicted in Figure 7 are virtually indistinguishable, with a near one-to-one agreement in peak position, providing convincing evidence that they are the same phase or crystal structure [56,57]. A comparison of the calculated PXRD profile for the co-crystal and the experimental OPT-NCC profile is depicted in Figure S4 in the Supplementary Information.

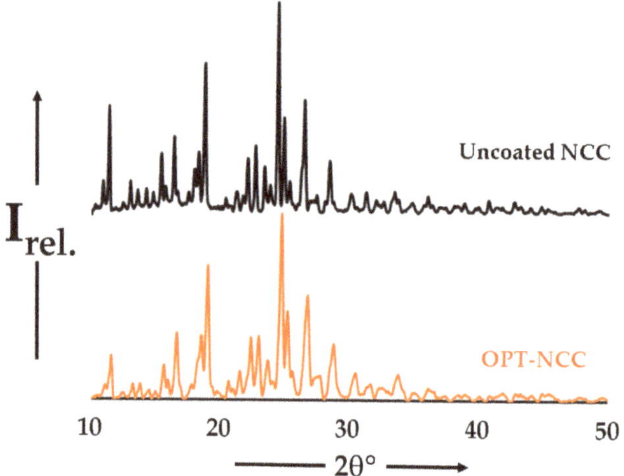

Figure 7. PXRD diffractograms for the uncoated (black) and OPT-NCC (orange).

3.2.5. Transmission Electron Microscopy (TEM)

The TEM micrograph depicted in Figure 8 reveals the crystals produced were prismatic or plate-like. This differs from the observations made by Zhang et al., that nanocrystals grown at low temperature are usually rod shaped [58]. Additional TEM images (Figures S5–S7) and a size distribution plot (Figure S8) have been included in the Supplementary Information.

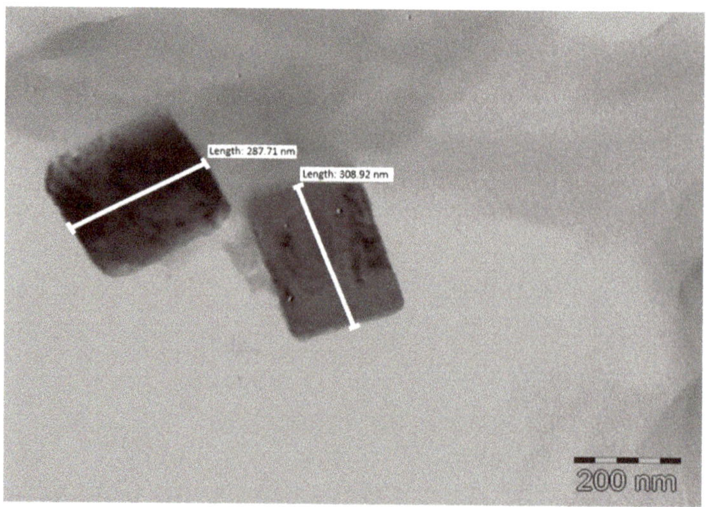

Figure 8. TEM micrograph of prismatic OPT-NCC.

3.2.6. Cytotoxicity Studies

The OPT-NCC exhibited improved cell viability when compared to that observed for the raw materials alone, which may be due to the stabilization induced by surfactants, and could have a shielding effect on the NCC. In addition, the presence of PEG, which forms the hydrophilic heads of TPGS 1000 may offer stealth properties to the NCC, minimizing uptake by HeLa cells [59,60]. Macrophages preferentially target negatively charged particles that are <1000 nm in size [61–63]. As the NCC produced in these studies are negatively charged, are <1000 nm in dimension and exhibit low HeLa cell toxicity, the NCC have the potential to target HIV harboring macrophages passively without affecting non-phagocytic cells. The summary of the in vitro cell viability is listed in Table 10 and depicted in Figure 9.

Table 10. Summary of in vitro cytotoxicity results.

Compound	Cell Viability %	SD
OPT-NCC	76.9	5.0
Physical Mixture	52.6	5.1
AZT	49.6	6.2
3TC	62.3	4.7

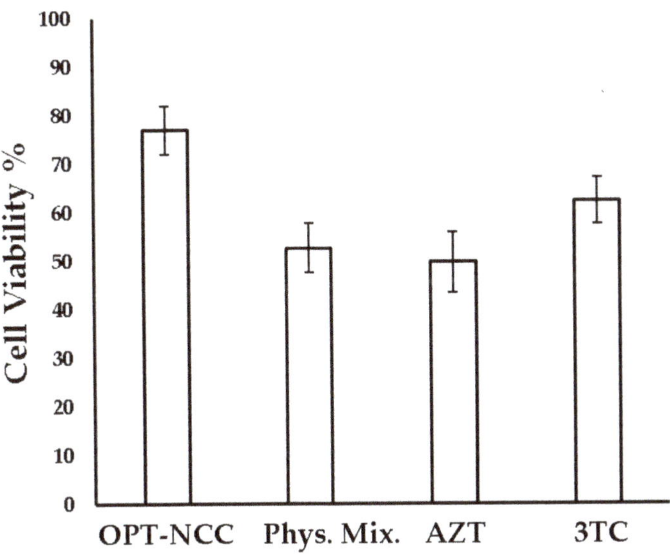

Figure 9. Cytotoxicity of NCC, individual API and physical mixture of active pharmaceutical ingredient (API).

4. Conclusions

The formulation optimization for production of electrosterically stabilized 3TC and AZT NCC using a pseudo one-solvent bottom-up approach was successful. The PS and PDI were affected by the amount of surfactant used in the formulation. The presence of surfactants reduces the energy at the surface of nucleating crystals preventing crystal growth [64]. The PS was not as affected by the concentration of TPGS 1000 used than when SDS was used suggesting that electrostatic stabilisation was more effective in the synthesis of the NCC using a bottom-up approach. The PDI was significantly affected by the concentration of SDS used, due to the combined effect of increasing SDS and TPGS 1000 concentrations in the composition.

The ZP is primarily dependent on the presence of SDS and a linear relationship exists between SDS concentration and ZP. The data reveal that an increase in SDS content results in a proportional decrease in ZP for the NCC suspensions produced.

In order to achieve the development of a suitable technology for the delivery of AZT and 3TC, the optimization objective was to minimize the ZP, increase the stability of the formulation [65,66] and with particle size reduction, potentially target macrophages [61–63,67].

Characterisation of the OPT-NCC formulation using TEM revealed that the NCC were <500 nm. PXRD and DSC data indicate the OPT-NCC are crystalline, which could potentially increase stability, and co-crystal formation was confirmed using FTIR. SEM-EDX was used to determine the elemental surface composition of the NCC. The spectra confirmed the presence of stabilizer on the surface of crystals in addition to the presence of molecular peaks associated with the API.

In vitro cytotoxicity studies revealed the OPT-NCC were less cytotoxic to HeLa cells than the individual API and a physical mixture of the API. The presence of hydrophilic PEG heads in TPGS 1000 may be the reason for increased HeLa cell viability.

The use of DoE and cold sonoprecipitation to manufacture optimized NCC is an inexpensive, reproducible and precise method of manufacturing NCC with specific pre-defined desirable PS, PDI and ZP, while maintaining crystallinity of the individual compounds. The NCC produced in this study exhibit less toxicity to HeLa cells than individual components, which in turn, may result in reduced side effects of each API. The improvement of the API side effect profile may well improve patient adherence to therapy and lead to better treatment outcomes.

Supplementary Materials: The following are available online at http://www.mdpi.com/1999-4923/12/4/367/s1, Table S1: Summary of melting temperatures of AZT, 3TC, reported co-crystal value and the OPT-NCC. Figure S1: DSC thermogram of 3TC (Black), AZT (Orange) and OPT-NCC (Blue). Table S2: Summary of the comparison FTIR for the reported co-crystal and the OPT-NCC. Figure S2: A plot of the values reported in Table S2 reflecting a one-to-one agreement between the FTIR wavenumbers reported in the literature [68] (blue) and those recorded for the NCC (orange). Figure S3: FTIR spectra of the OPT-NCC (orange), 3TC (blue) and AZT (black). Figure S4: PXRD diffractograms of OPT-NCC and Co-crystal (Diffractogram obtained from CSD [69]). Figure S5: TEM of uncoated bottom-up micro co-crystal. Figure S6: TEM depicting the mean particle size of the OPT-NCC, Figure S7: TEM images of the smallest product obtained (OPT-NCC). Figure S8: DLS intensity size distribution curve of the OPT-NCC.

Author Contributions: R.B.W. conceptualized, supervised and contributed to writing and editing of the manuscript. B.A.W. performed the experiments, analyzed the data and wrote the manuscript. V.J.S. contributed to the conceptualization, supervision, writing, editing, bibliography research and proofreading of the manuscript. All authors have read and agreed to the published version of the manuscript.

Funding: This research was not funded with an external research grant.

Acknowledgments: The authors acknowledge the National Research Foundation (B.A.W.) and the Research Committee of Rhodes University (R.B.W.) and Rhodes University Sandisa Imbewu fund (V.J.S.) for financial assistance. The contribution of Pascal Ntemi for running some samples is gratefully acknowledged.

Conflicts of Interest: The authors declare no conflict of interest.

References

1. UNAIDS. *Global HIV/AIDS Statistics—2019 Fact Sheet*; UNAIDS: Geneva, Switzerland, 2019.

2. Sekhon, B. Pharmaceutical co-crystals—A review. *ARS Pharm.* **2009**, *150*, 99–117.

3. Brittain, H.G. Cocrystal Systems of Pharmaceutical Interest: 2010. *Cryst. Growth Des.* **2011**, *36*, 361–381. [CrossRef]

4. Gao, Y.; Zu, H.; Zhang, J. Enhanced dissolution and stability of adefovir dipivoxil by cocrystal formation. *J. Pharm. Pharmacol.* **2011**, *63*, 483–490. [CrossRef] [PubMed]

5. Bethune, S.J.; Schultheiss, N.; Henck, J.O. Improving the poor aqueous solubility of nutraceutical compound pterostilbene through cocrystal formation. *Cryst. Growth Des.* **2011**, *11*, 2817–2823. [CrossRef]

6. Jayasankar, A.; Reddy, L.S.; Bethune, S.J.; Rodríguez-Hornedo, N. Role of Cocrystal and Solution Chemistry on the Formation and Stability of Cocrystals with Different Stoichiometry. *Cryst. Growth Des.* **2009**, *9*, 889–897. [CrossRef]

7.   Sekhon, B.S. Drug-drug co-crystals. *DARU J. Pharm. Sci.* **2012**, *20*. [CrossRef] [PubMed]

8.   De Clercq, E. Antiretroviral drugs. *Curr. Opin. Pharmacol.* **2010**, *10*, 507–515. [CrossRef]

9.   Breckenridge, A. Pharmacology of drugs for HIV. *Medicine* **2009**, *37*, 374–377. [CrossRef]

10. Tsibris, A.M.N.; Hirsch, M.S. Antiretroviral Therapy for Human Immunodeficiency Virus Infection. In *Mandell, Douglas, and Bennett's Principles and Practice of Infectious Diseases*; Bennett, J.E., Dolin, R., Blaser Douglas, and Bennett's Principles and Practice of Infectious Diseases (Eighth Edition), M.J.B.T.-M., Eds.; Elsevier: Philadelphia, PA, USA, 2014; Volume 1, pp. 1622–1641.e6, ISBN 9996096742.

11. Wainberg, M.A. AIDS: Drugs that prevent HIV infection. *Nature* **2011**, *469*, 306–307. [CrossRef]

12. Bhatt, P.M.; Azim, Y.; Thakur, T.S.; Desiraju, G.R. Co-Crystals of the Anti-HIV Drugs Lamivudine and Zidovudine. *Cryst. Growth Des.* **2009**, *9*, 951–957. [CrossRef]

13. Xia, D.; Gan, Y.; Cui, F. Application of Precipitation Methods for the Production of Water-insoluble Drug Nanocrystals: Production Techniques and Stability of Nanocrystals. *Curr. Pharm. Des.* **2014**, *20*, 408–435. [CrossRef] [PubMed]

14. Dalvi, S.V.; Yadav, M.D. Effect of ultrasound and stabilizers on nucleation kinetics of curcumin during liquid antisolvent precipitation. *Ultrason. Sonochem.* **2015**, *24*, 114–122. [CrossRef] [PubMed]

15. Shegokar, R.; Müller, R.H. Nanocrystals: Industrially feasible multifunctional formulation technology for poorly soluble actives. *Int. J. Pharm.* **2010**, *399*, 129–139. [CrossRef] [PubMed]

16. Merisko-Liversidge, E.; Liversidge, G.G.; Cooper, E.R. Nanosizing: A formulation approach for poorly-water-soluble compounds. *Eur. J. Pharm. Sci.* **2003**, *18*, 113–120. [CrossRef]

17. Pi, J.; Liu, Z.; Wang, H.; Gu, X.; Wang, S.; Zhang, B.; Luan, H.; Zhu, Z. Ursolic Acid Nanocrystals for Dissolution Rate and Bioavailability Enhancement: Influence of Different Particle Size. *Curr. Drug Deliv.* **2016**, *13*, 1358–1366. [CrossRef]

18. Nakarani, M.; Patel, P.; Patel, J.; Patel, P.; Murthy, R.S.R.; Vaghani, S.S. Cyclosporine A-Nanosuspension: Formulation, characterization and in vivo comparison with a marketed formulation. *Sci. Pharm.* **2010**, *78*, 345–361. [CrossRef]

19. Cerdeira, A.M.; Mazzotti, M.; Gander, B. Pharmaceutical Nanotechnology Miconazole nanosuspensions: Influence of formulation variables on particle size reduction and physical stability. *Int. J. Pharm.* **2010**, *396*, 210–218. [CrossRef]

20. Iurian, S.; Bogdan, C.; Tomuță, I.; Szabó-Révész, P.; Chvatal, A.; Leucuța, S.E.; Moldovan, M.; Ambrus, R. Development of oral lyophilisates containing meloxicam nanocrystals using QbD approach. *Eur. J. Pharm. Sci.* **2017**, *104*, 356–365. [CrossRef]

21. Chow, S.F.; Wan, K.Y.; Cheng, K.K.; Wong, K.W.; Sun, C.C.; Baum, L.; Chow, A.H.L. Development of highly stabilized curcumin nanoparticles by flash nanoprecipitation and lyophilization. *Eur. J. Pharm. Biopharm.* **2015**, *94*, 436–449. [CrossRef]

22. Dong, Y.; Ng, W.K.; Shen, S.; Kim, S.; Tan, R.B.H. Preparation and characterization of spironolactone nanoparticles by antisolvent precipitation. *Int. J. Pharm.* **2009**, *375*, 84–88. [CrossRef]

23. Leng, D.; Chen, H.; Li, G.; Guo, M.; Zhu, Z.; Xu, L.; Wang, Y. Development and comparison of intramuscularly long-acting paliperidone palmitate nanosuspensions with different particle size. *Int. J. Pharm.* **2014**, *472*, 380–385. [CrossRef]

24. De Smet, L.; Saerens, L.; De Beer, T.; Carleer, R.; Adriaensens, P.; Van Bocxlaer, J.; Vervaet, C.; Remon, J.P. Formulation of itraconazole nanococrystals and evaluation of their bioavailability in dogs. *Eur. J. Pharm. Biopharm.* **2014**, *87*, 107–113. [CrossRef]

25. Witika, B.A.; Smith, V.J.; Walker, R.B. A Comparative Study of the Effect of Different Stabilizers on the Critical Quality Attributes of Self- assembling Nano Co-crystals. *Pharmaceutics* **2020**, *12*, 182. [CrossRef] [PubMed]

26. Nowacek, A.S.; Mcmillan, J.; Miller, R.; Anderson, A.; Rabinow, B.; Gendelman, H.E. Macrophages: Implications for NeuroAIDS Therapeutics. *J. Neuroimmune Pharmacol.* **2012**, *5*, 592–601. [CrossRef] [PubMed]

27. Orenstein, J.M. Replication of HIV-1 in vivo and in vitro. *Ultrastruct. Pathol.* **2007**, *31*, 151–167. [CrossRef] [PubMed]

28. Koppensteiner, H.; Wu, L. Macrophages and their relevance in Human Immunodeficiency Virus Type I infection. *Retrovirology* **2012**, *9*, 82. [CrossRef]

29. Van't Klooster, G.; Hoeben, E.; Borghys, H.; Looszova, A.; Bouche, M.-P.P.; Van Velsen, F.; Baert, L. Pharmacokinetics and disposition of rilpivirine (TMC278) nanosuspension as a long-acting injectable antiretroviral formulation. *Antimicrob. Agents Chemother.* **2010**, *54*, 2042–2050. [CrossRef]

30. Kelly, C.; Jefferies, C.; Cryan, S.-A. Targeted Liposomal Drug Delivery to Monocytes and Macrophages. *J. Drug Deliv.* **2011**, *2011*, 727241. [CrossRef]
31. Bender, A.; Schäfer, V.; Steffan, A.M.; Royer, C.; Kreuter, J.; Rübsamen-Waigmann, H.; von Briesen, H. Inhibition of HIV in vitro by antiviral drug-targeting using nanoparticles. *Res. Virol.* **1994**, *145*, 215–220. [CrossRef]
32. Désormeaux, A.; Bergeron, M.G. Lymphoid tissue targeting of anti-HIV drugs using liposomes. *Methods Enzymol.* **2005**, *391*, 330–351.
33. Bučar, D.-K.; MacGillivray, L.R. Preparation and Reactivity of Nanocrystalline Cocrystals Formed via Sonocrystallization. *J. Am. Chem. Soc.* **2007**, *129*, 32–33. [CrossRef] [PubMed]
34. MacGillivray, L.R.; Sander, J.R.; Bucar, D.K.; Elacqua, E.; Zhang, G.; Henry, R. Sonochemical synthesis of nano-cocrystals. *Proc. Mtgs. Acoust.* **2013**, *19*, 45090.
35. Pawar, V.K.; Singh, Y.; Meher, J.G.; Gupta, S.; Chourasia, M.K. Engineered nanocrystal technology: In-vivo fate, targeting and applications in drug delivery. *J. Control. Release* **2014**, *183*, 51–66. [CrossRef] [PubMed]
36. Lu, Y.; Li, Y.; Wu, W. Injected nanocrystals for targeted drug delivery. *Acta Pharm. Sin. B* **2016**, *6*, 106–113. [CrossRef] [PubMed]
37. Sathisaran, I.; Dalvi, S.V. Engineering cocrystals of poorly water-soluble drugs to enhance dissolution in aqueous medium. *Pharmaceutics* **2018**, *10*, 108. [CrossRef]
38. Brittain, H.G. Pharmaceutical cocrystals: The coming wave of new drug substances. *J. Pharm. Sci.* **2013**, *102*, 311–317. [CrossRef]
39. Desai, P.P.; Patravale, V.B. Curcumin Cocrystal Micelles—Multifunctional Nanocomposites for Management of Neurodegenerative Ailments. *J. Pharm. Sci.* **2018**, *107*, 1143–1156. [CrossRef]
40. Bezerra, M.A.; Santelli, R.E.; Oliveira, E.P.; Villar, L.S.; Escaleira, L.A. Response surface methodology (RSM) as a tool for optimization in analytical chemistry. *Talanta* **2008**, *76*, 965–977. [CrossRef] [PubMed]
41. Wong, H.L.; Chattopadhyay, N.; Wu, X.Y.; Bendayan, R. Nanotechnology applications for improved delivery of antiretroviral drugs to the brain. *Adv. Drug Deliv. Rev.* **2010**, *62*, 503–517. [CrossRef]
42. Hill, W.J.; Hunter, W.G. A Review of Response Surface Methodology: A Literature Review. *Technometrics* **2012**, *8*, 571–590. [CrossRef]
43. Matteucci, M.E.; Hotze, M.A.; Johnston, K.P.; Williams, R.O. Drug nanoparticles by antisolvent precipitation: Mixing energy versus surfactant stabilization. *Langmuir* **2006**, *22*, 8951–8959. [CrossRef] [PubMed]
44. Quan, P.; Shi, K.; Piao, H.; Piao, H.; Liang, N.; Xia, D.; Cui, F. A novel surface modified nitrendipine nanocrystals with enhancement of bioavailability and stability. *Int. J. Pharm.* **2012**, *430*, 366–371. [CrossRef] [PubMed]
45. Zimmermann, A.; Millqvist-Fureby, A.; Elema, M.R.; Hansen, T.; Müllertz, A.; Hovgaard, L. Adsorption of pharmaceutical excipients onto microcrystals of siramesine hydrochloride: Effects on physicochemical properties. *Eur. J. Pharm. Biopharm.* **2009**, *71*, 109–116. [CrossRef] [PubMed]
46. Keck, C.M. Cyclosporine Nanosuspensions: Optimised Size Characterisation and Oral Formulations. Ph.D. Thesis, Freie Universitat, Berlin, Germany, 2006.
47. Hou, Y.; Shao, J.; Fu, Q.; Li, J.; Sun, J.; He, Z. Spray-dried nanocrystals for a highly hydrophobic drug: Increased drug loading, enhanced redispersity, and improved oral bioavailability. *Int. J. Pharm.* **2017**, *516*, 372–379. [CrossRef]
48. Rachmawati, H.; Rahma, A.; Al Shaal, L.; Müller, R.H.; Keck, C.M. Destabilization mechanism of ionic surfactant on curcumin nanocrystal against electrolytes. *Sci. Pharm.* **2016**, *84*, 685–693. [CrossRef]
49. Dzakwan, M.; Pramukantoro, G.E.; Mauludin, R.; Wikarsa, S. Formulation and characterization of fisetin nanosuspension. In *IOP Conference Series: Materials Science and Engineering*; IOP: London, UK, 2017; Volume 259.
50. Ige, P.P.; Baria, R.K.; Gattani, S.G. Fabrication of fenofibrate nanocrystals by probe sonication method for enhancement of dissolution rate and oral bioavailability. *Colloids Surfaces B Biointerfaces* **2013**, *108*, 366–373. [CrossRef]
51. Lopeandía, A.F.; Rodríguez-Viejo, J. Size-dependent melting and supercooling of Ge nanoparticles embedded in a SiO2 thin film. *Thermochim. Acta* **2007**, *461*, 82–87. [CrossRef]
52. Sun, J.; Simon, S.L. The melting behavior of aluminum nanoparticles. *Thermochim. Acta* **2007**, *463*, 32–40. [CrossRef]

53. Palermo, E.F.; Chiu, J. Critical review of methods for the determination of purity by differential scanning calorimetry. *Thermochim. Acta* **1976**, *14*, 1–12. [CrossRef]

54. Brink, G. Infrared Studies of Water in Crystalline Hydrates: Ba(ClO3)2·H2O. *Appl. Spectrosc.* **1976**, *30*, 630–631. [CrossRef]

55. Falk, M.; Huang, C.-H.; Knop, O. Infrared Spectra of Water in Crystalline Hydrates: KSnCl3.H2O, an Untypical Monohydrate. *Can. J. Chem.* **2006**, *52*, 2928–2931. [CrossRef]

56. Kamb, W.B. Theory of Preferred Crystal Orientation Developed. *J. Geol.* **1958**, *67*, 153–170. [CrossRef]

57. Lang, A.R. X-ray diffraction procedures for polycrystal-line and amorphous materials. *Acta Metall.* **1956**, *4*, 102. [CrossRef]

58. Zhang, X.; Xia, Q.; Gu, N. Preparation of all-trans retinoic acid nanosuspensions using a modified precipitation method. *Drug Dev. Ind. Pharm.* **2006**, *32*, 857–863. [CrossRef] [PubMed]

59. Romberg, B.; Hennink, W.E.; Storm, G. Sheddable coatings for long-circulating nanoparticles. *Pharm. Res.* **2008**, *25*, 55–71. [CrossRef]

60. Jadhav, P.; Bothiraja, C.; Pawar, A. Methotrexate-Loaded Nanomixed Micelles: Formulation, Characterization, Bioavailability, Safety, and In Vitro Anticancer Study. *J. Pharm. Innov.* **2018**, *13*, 213–225. [CrossRef]

61. Chono, S.; Tanino, T.; Seki, T.; Morimoto, K. Uptake characteristics of liposomes by rat alveolar macrophages: Influence of particle size and surface mannose modification. *J. Pharm. Pharmacol.* **2007**, *59*, 75–80. [CrossRef]

62. Chono, S.; Tauchi, Y.; Morimoto, K. Influence of particle size on the distributions of liposomes to atherosclerotic lesions in mice. *Drug Dev. Ind. Pharm.* **2006**, *32*, 125–135. [CrossRef]

63. Chono, S.; Tanino, T.; Seki, T.; Morimoto, K. Influence of particle size on drug delivery to rat alveolar macrophages following pulmonary administration of ciprofloxacin incorporated into liposomes. *J. Drug Target.* **2006**, *14*, 557–566. [CrossRef]

64. Rabinow, B.E. Nanosuspensions in drug delivery. *Nat. Rev. Drug Discov.* **2004**, *3*, 785. [CrossRef]

65. Müller, R.H.; Jacobs, C. Buparvaquone mucoadhesive nanosuspension: Preparation, optimisation and long-term stability. *Int. J. Pharm.* **2002**, *237*, 151–161. [CrossRef]

66. Sutradhar, K.B.; Khatun, S.; Luna, I.P. Increasing possibilities of nanosuspension. *J. Nanotechnol.* **2013**, *2013*. [CrossRef]

67. Ahsan, F.; Rivas, I.P.; Khan, M.A.; Torres Suárez, A.I. Targeting to macrophages: Role of physicochemical properties of particulate carriers—Liposomes and microspheres—On the phagocytosis by macrophages. *J. Control. Release* **2002**, *79*, 29–40. [CrossRef]

68. Lupin Limited. *WO 2009/116055 Al 2009*; Lupin Limited: Mumbaid, India, 2009.

69. Groom, C.R.; Bruno, I.J.; Lightfoot, M.P.; Ward, S.C. The Cambridge Structural Database. *Acta Crystallogr. Sect. B Struct. Sci. Cryst. Eng. Mater.* **2016**, *72*, 171–179. [CrossRef] [PubMed]

MDPI

St. Alban-Anlage 66

4052 Basel

Switzerland

Tel. +41 61 683 77 34

Fax +41 61 302 89 18

www.mdpi.com

Pharmaceutics Editorial Office

E-mail: pharmaceutics@mdpi.com

www.mdpi.com/journal/pharmaceutics